KU-097-564

Digital Jacquard Design

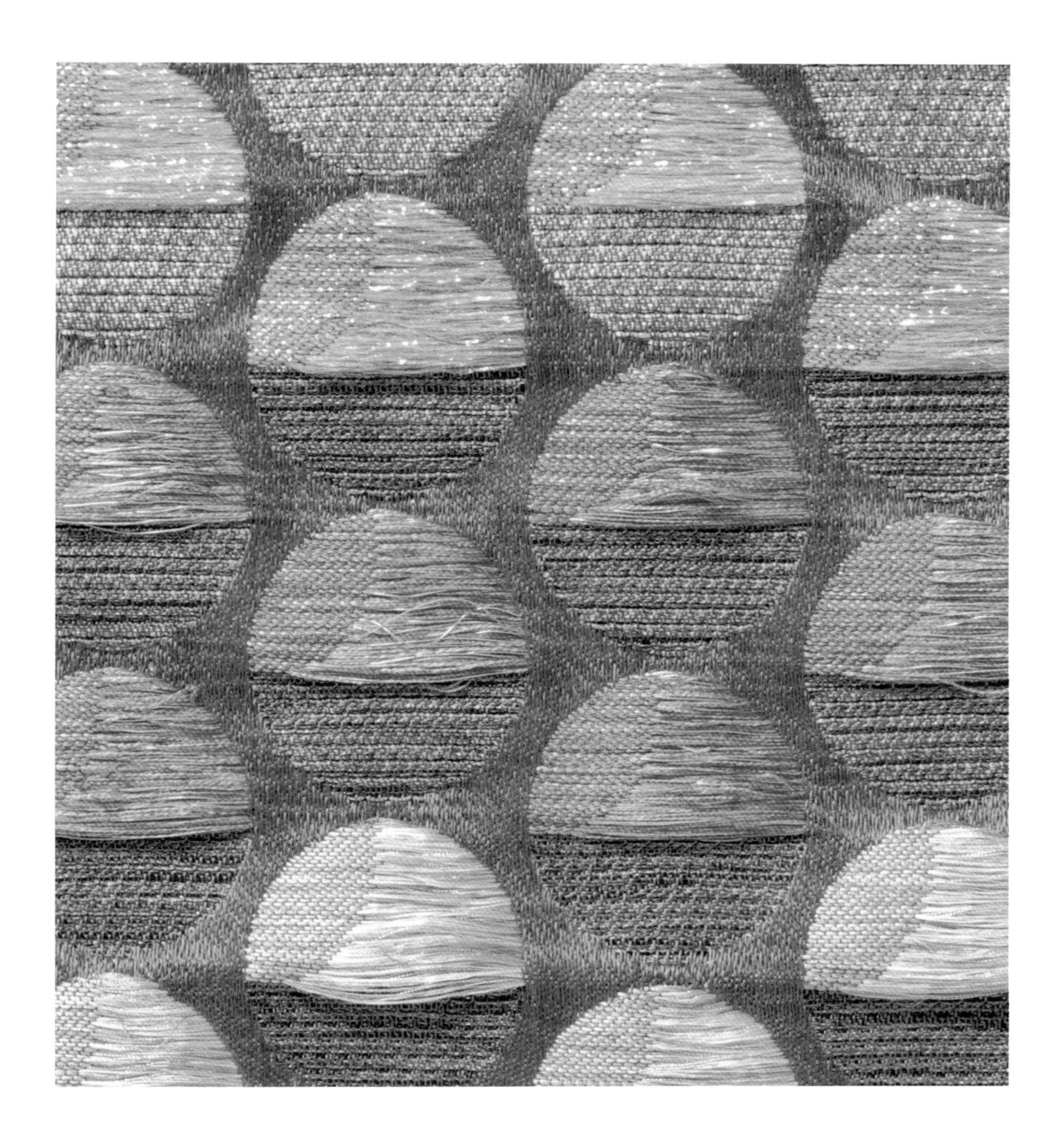

Happles, *handwoven figured silk with light reflective pattern weft by Melanie Olde, LFS.*

To the Student Who Asks Questions

and of course, to John and Dario, who made it possible

This page: Leaf, *a self-patterned silk damask by Melanie Olde, LFS.*

Facing page: Green Triangles, *a brocaded silk damask by Elisabeth Egger, LFS.*

Digital Jacquard Design

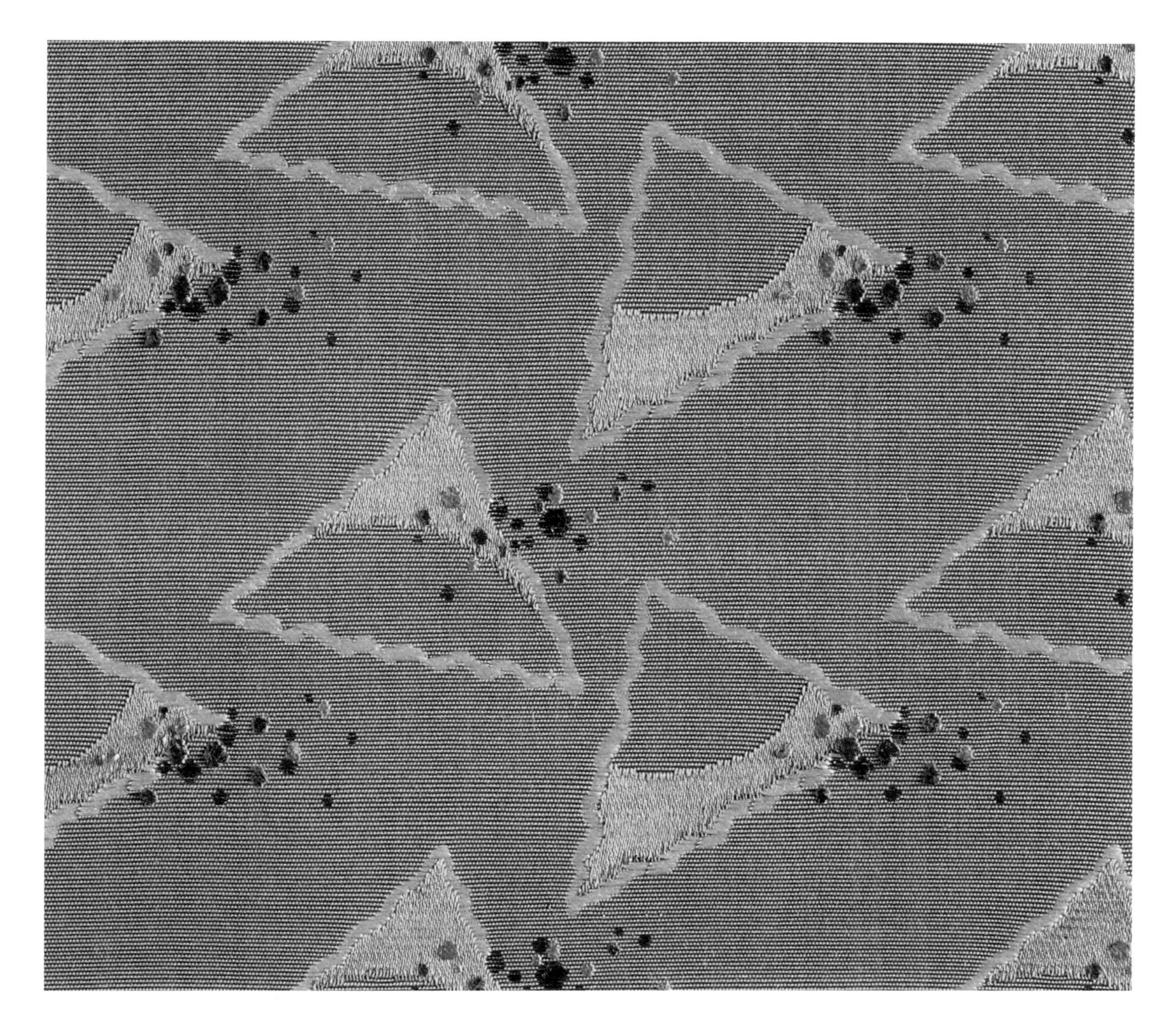

Julie Holyoke

including photographs and illustrations by Dario Bartolini

BLOOMSBURY
LONDON • NEW DELHI • NEW YORK • SYDNEY

Bloomsbury Academic

An imprint of Bloomsbury Publishing Plc

50 Bedford Square
London
WC1B 3DP
UK

1385 Broadway
New York
NY 10018
USA

www.bloomsbury.com

Bloomsbury is a registered trade mark of Bloomsbury Publishing Plc

First published 2013

British Library Cataloguing-in-Publication Data
A catalogue record for this book is available from the British Library.

ISBN: HB: 978-0-85785-345-5

Library of Congress Cataloging-in-Publication Data

Holyoke, Julie.
Digital jacquard design / Julie Holyoke. — First edition.
pages cm
Includes bibliographical references and index.
ISBN 978-0-85785-345-5 (hardback)
1. Textile design. 2. Textile fabrics. 3. Jacquard weaving. I. Title.
TS1475.H76 2013
746—dc23
2013020290

Typeset by Apex CoVantage, LLC, Madison, WI, USA
Printed and bound in China

Page Designer: Susan McIntyre

Contents

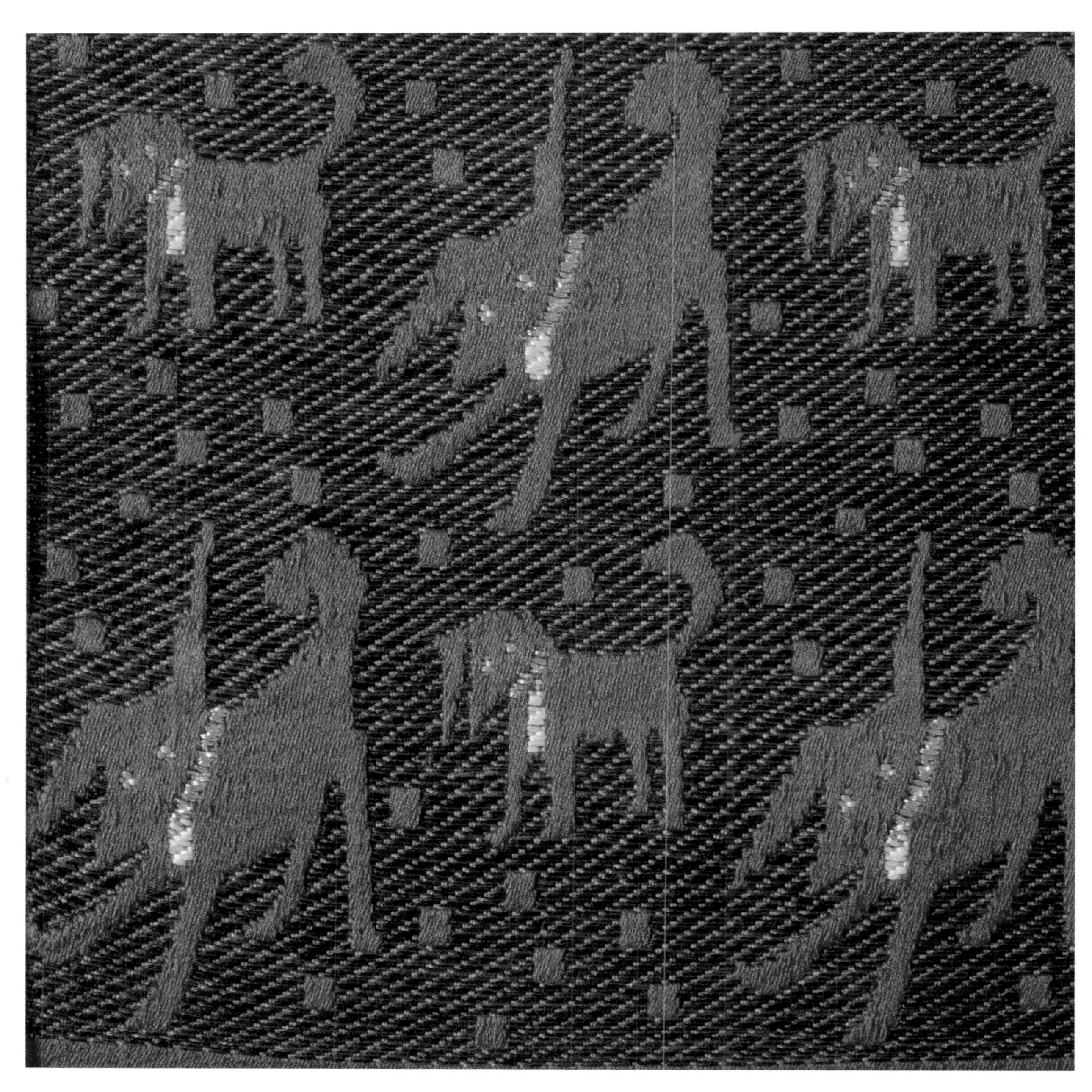

Figure 1. Briciola at Night, *lampas with silk and metalic wefts by Barbara Shawcroft, LFS.*

Acknowledgements

The Lisio Foundation in Florence, Italy made this book possible by generously granting permission to reproduce a selection of figured textiles from its archives. The majority of the textiles seen in these pages come from an extraordinary collection of student projects, woven by the gifted and dedicated scholars of figured textile design that have attended the foundation's courses for over two decades. To the Lisio Foundation School and all those who have donated their work to the archives for others' instruction, my infinite thanks.

During the course of the two years this volume took to prepare, a number of these students sent additional samples and documentation of their projects and helped me to renew contacts that had lapsed over time. Recognition is gratefully given for their support; their names follow in alphabetical order: Martin Ciszuk, Berthe Forchammer, Rudy Kovacs, Tuulia Lampinen, Helena Loermans, Tina Moor, Antoinette Stucky, Barbara Shawcroft, Tanja Valta, Bhakti Ziek.

I would also like to thank the president of the foundation and four colleagues for their individual contributions to this project: Francesco Ortona, Paola Marabelli, Marta Valdarni, Oriana Castagnozzi, and Cecilia Cerchiarini.

Artists, designers, and weavers have contributed images, text, and technical documentation of textiles woven at mills and textile programs in Europe and North America. To them a special thanks for their generosity and willingness to share an inside view of their works with others: Catharine Ellis, Bethanne Knudson, Pirita Lauri, Janice-Lessman Moss, Robin Muller, Elaine Yan Ling Ng, and Bhakti Ziek. Philippa Brock and Ruth Scheuing, whose works are not (regretfully) pictured on these pages, helped this book come into being.

A number of historical textiles enrich the contents of this text. The inclusion of these artifacts was made possible by the kindness of the Cleveland Museum of Art; Titi Halle at Cora Ginsburg LLC; Severine Experton-Dard, Georges Le Manach, and Thierry Maigret; Paola Marabelli of the Lisio Foundation; Marilena de Vecchi Ranieri; and John Marshall.

For help and suggestions on classification, terminology, and technology, I am indebted to Martin Ciszuk, Titi Halle, Michele Majer, Paola Marabelli, Bhakti Ziek, Bethanne Knudson, Kelly Hopkin, and Silvano Mazzoni.

I apologize for any inadvertent errors or omissions that may have occurred when titling, crediting, drafting, or describing the wonderful works of so many gifted colleagues and students, and once again thank you all.

Special thanks are due to Rita Comanducci, whose knowledge of the economics of silk weaving in the Renaissance has enlightened students and colleagues over the years, and whose expertise informed the affirmation with which this book begins.

For generous gifts of expertise and advice, thank you, Nancy Holyoke, Lydie Hudson, and John Millerchip.

I am indebted to Geraldine Billingham for the tactful phrasing of her replies and for her infinite patience.

Abbreviations

CJH	Author's collection
LFAS	Lisio Foundation Archivio Storico
LFCHT	Lisio Foundation Collection of Historical Textiles
LFS	Lisio Foundation School

Introduction

During the Italian Renaissance, figured textiles were regarded as works of art, fit for princely garb and occasions of state. The amount paid for a *braccio*[1] of figured cloth reflected more than labor and material costs; it represented the value attributed to the uniqueness of such artifacts. At the beginning of the nineteenth century, the invention of Jacquard's machine ensured the repeatability of a design, and figured cloth took a plunge from rare to reproducible goods. Over the centuries, other creative forms supplanted figured wovens, now dubbed Jacquard, as gifts of state and works of art.

As cloth is woven, one of two things occurs: warp ends are raised or remain stationary when the weft traverses the loom's width. In Jacquard weaving, warp ends are raised selectively to create the weave-patterning that distinguishes figured textiles. For as long as such textiles have been woven, the transmission of this data, the warp's position relative to the weft, has remained a complex and time-consuming process.

The advent of digital technology, so ideally suited to a medium that requires a program for one of two options – raised or stationary warp end, has radically changed the tools and processes of figured weaving. Until a few decades ago, a series of specialists was required to complete the many steps from first sketch to finished bolt of intricately woven cloth. With today's tools, one skilled designer/technician can control the entire cycle, changing the skill sets required of those who work in industry. Not only has industrial practice changed: new digital tools have made figuring technology available to artists and handweavers. Today lighter, less cumbersome looms are commonplace in university art departments and individual studios. After centuries, figured weaving has returned to take its place among the major arts, be it in the form of industrially woven tapestry or handwoven intelligent textile.

The publication of texts that meet the educational needs of artists and designers has lagged behind changes in the medium. *Digital Jacquard Design* fills this void and provides both aesthetic and technical training for today's students of figured weaving.

Figure II. Facing page: *Japanese silk velvet, courtesy of John Marshall.*

Part I

The Designer's Skill Set

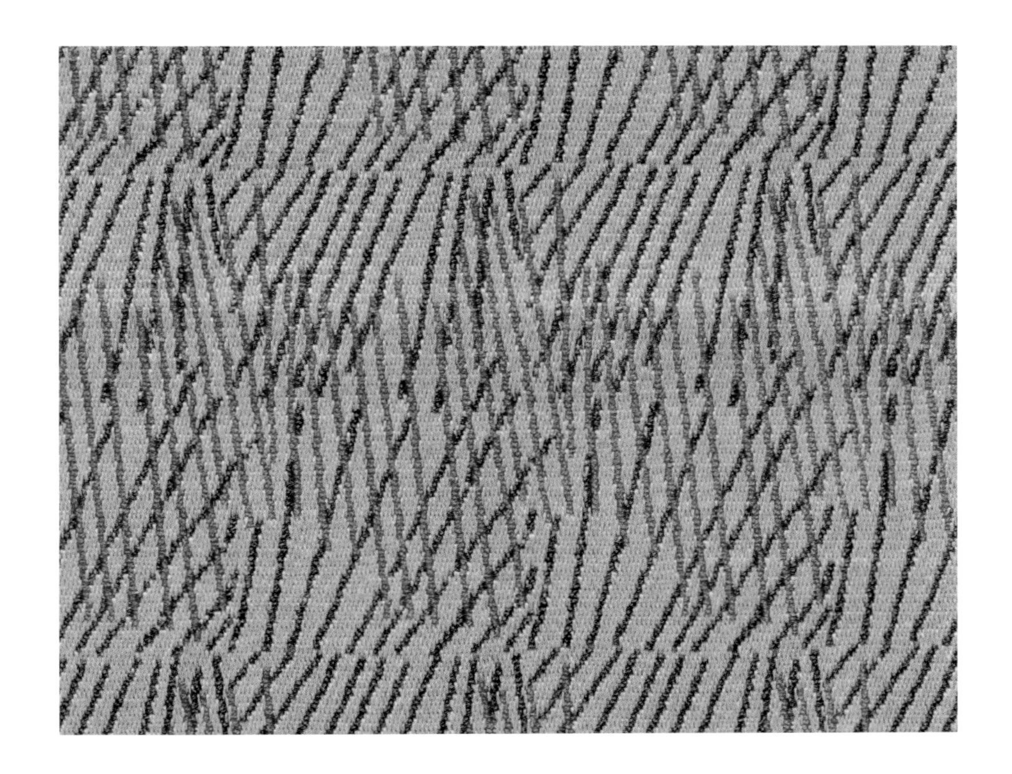

Sticks *by Tuulia Lampinen, LFS.*

1 Visual analysis: identifying contrasts

What makes a weave-patterned textile visually effective? Why does it draw our attention, intrigue or please us? Why do we ignore it, walk up to it for a better look, want to touch it, or move away? What factors contribute to making a figured textile "work"?

Through observation and visual analysis, a designer gains an understanding of how structure, material, and design work in combination to produce the two or more distinct woven surfaces of a figured textile. The differences between each of these surfaces, or weave effects, can be defined in terms of **type, degree, and quality** of contrast.

A first group of damasks illustrates how three factors, classified as types of contrast, generate visible differences on the surface of a textile. These contrasts can be further defined in terms of degree and quality.

Type of contrast

Structural

Weave class: tabby and derivatives, twill, or satin
Warp, weft-faced, or balanced weaves
Simple or compound weaves
Warp and weft setts

Material

Fiber, yarn construction, color

Design-Based

Number of effects
Area of each effect
Dimension and form of motifs
Repeat

Color contrast is virtually absent in the three damasks shown in Figure 1.1. In the lowermost fabric, warp and weft-faced satins 5, and differences between the longer, parallel fibers of the warp and the shorter, softer fibers of the weft, produce a minimal contrast of light reflection that gives an evanescent quality to the patterning.

Figure 1.1. Facing page: From top to bottom, Triangles *by Annette Schyren, LFS;* Leaf Skeleton *by Janina von Weissenberg, LFS; linen damask, CJH.*

Figure 1.2. Right: *Eighteenth-century silk damask with areas of small, geometric patterning. Courtesy of Cora Ginsburg LLC.*

Patterning in *Leaf Skeleton*, the damask immediately above, is the result of structural and material differences. A silk organzine warp bound in warp satin 8 reflects light, while an opaque cotton weft and 2/2 basket break the reflective properties of the silk to create a matt surface.

In the uppermost damask, contrast is the product of a combination of structure, materials, and design. Warp and weft-faced satins 8 are woven on a silk warp identical to that of *Leaf Skeleton*, but with a spongy wool weft. The different elasticity of warp and weft, as well as the frequent exchange from warp to weft-faced satin along the horizontal lines of the design, generate pronounced ridges.

The cherry-colored *Bizarre* silk[1] in Figure 1.3, woven with warp and weft-faced satins and tone-on-tone materials, is a traditional damask. Contrasts of form lend interest to this damask, as curvilinear, rounded fruits and flowers are juxtaposed to spiny leaves and fronds.

A network of streets and city blocks in *Florence Map* are translated into two weaves from different families, a smooth warp satin 8 and a pebbly plain weave derivative. The satin enhances, while the plain weave breaks up the light reflective quality of the silk warp; the tighter/looser interlacement adds further texture and light/dark contrast to the surface.

The striped warp (atypical in figured damask) that connotes woolen *Norwich damask* is woven with two weave structures from the same class: satin, in warp and weft-faced versions. Variations of warp color modulate the visibility of the weave patterning. Against the darker, solid warp stripes, the light-colored weft satin stands out clearly; against lighter colors, it recedes into the ground; in correspondence with the finer warp stripes, the weft-faced areas virtually disappear. Though not shown in the photograph, the widths of the weave repeat and warp stripe coincide, and do not produce an expansion of the overall width of the textile's design.

Figure 1.3. Left: *Eighteenth-century* Bizarre *silk, courtesy of Cora Ginsburg LLC;* Below: Florence Map *by Lindsay Lantz and Zoe Sargent, LFS.*

In the three damasks that follow, more than two weave effects are used, thus extending the potential for creating a range of effects, but also posing new challenges for the designer, who must choose how best to "spend" the limited amount of contrast afforded by the single warp and weft series that characterize damask.

Like the *Norwich damask*, geometrically patterned *Shaded Squares* adds the variable of material color to the interaction between structure, materials, and design. Seven distinct weaves (a series of shaded 8-end satins), together with a diminutive motif, create a small weave repeat equal to the size of one square of the woven textile. Bands of weft colors are arranged to coincide with the length of the motif/weave repeat. These variations in color, together with the shaded satins, produce six new colors in each horizontal row (the warp-faced satin remains a constant white). Each new weft color also increases the overall length of the textile's repeat, as the height of the weave repeat is multiplied by the total number of weft colors.

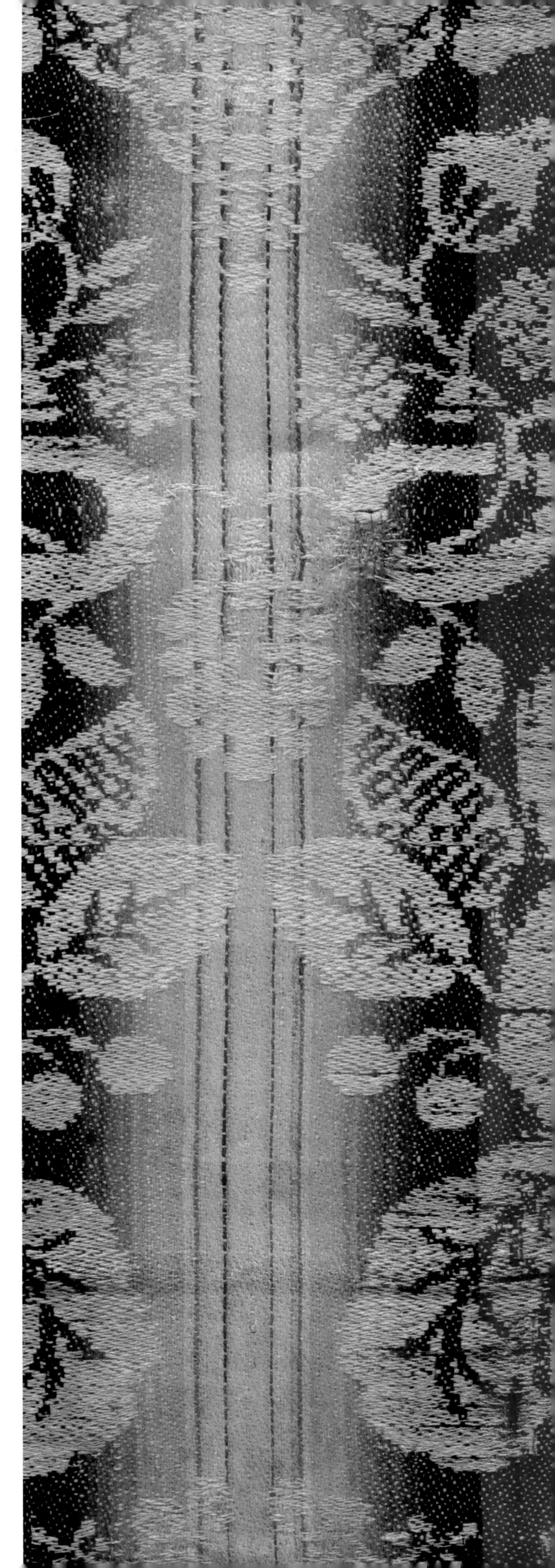

Figure 1.4. Right: Norwich damask, *courtesy of Cora Ginsburg LLC;* Below: Shaded Squares *by Helena Loermans, LFS.*

Two more damasks conclude this section on how structure, material, and design work in combination to produce the varied surfaces that are the visual language of weave-patterned textiles.

In *Rooftops*, a light warp and dark weft are woven with three structures: warp and weft satins, and a tighter plain weave derivative. Color and weave structure work in combination to produce two highly contrasted and one intermediate, low-contrast surface. *Nettles* is also woven with a light warp and dark weft that generate a strong value contrast, but here the interval between contrasts is reduced by the total number of weaves, all from the same family: seven 8-end shaded satins generate a gradual blending of warp-to-weft combinations and limit textural contrast to a minimum.

Figure 1.5. Left: Rooftops *by Sheetal Khanna-Ravich, LFS;* Below: Nettles *by Cecilia Crociani, LFS.*

Degree and quality of contrast

Many figuring techniques are structurally more complex than damask. Added complexity allows the designer more margin for creating distinct weave areas. Noting the degree of contrast, or the amount of difference between weave effects in existing textiles, is a useful exercise of visual analysis, as is the definition of the quality, or character, of contrast. When defining degree, opposites come to mind, such as high/low, strong/subtle, dramatic/discreet, but it should be kept in mind that weave patterning allows great freedom and intervals of contrast that may not be equidistant.

The quality of a contrast is often easiest to think of in paired opposites, such as: light/dark, subdued/brilliant, flat/raised, matt/glossy, smooth/rough. The design of a figured textile can be defined as: simple/complex, plain/elaborate, empty/full, and so forth.

In *Lace,* the structure of damask becomes more complex than that of *Onions,* a damask woven with two simple weaves. The raised lace network that overlies a damask ground is produced by the introduction of a compound weave that allows alternate picks of the ground weft to float above a tighter weave that continues below. More surface differentiation is achieved, but no extra weft or weaving time is needed. Silk warp and weft are close in color, but the opaque, stiff quality of a linen weft throws the floats of the surface picks into clear relief.

Figure 1.6. Above: Onions, *a simple damask by Hans Thomsson, LFS;* Right: Lace, *a damask with simple and compound weaves by Sara Thorn, LFS.*

The next two textiles use the same device: every other ground pick weaves not as a tightly bound ground weave, but is left to float either freely or for a longer interval above the ground to create more pronounced patterning. This structural device allows for a greater division between ground and patterning.

With the addition of compound structure and more materials, the potential for creating distinct woven surfaces increases, but as means augment so does complexity of choice. A selection of the opportunities for contrast that weave patterning offers is shown on the pages that follow. Each pair of facing pages explores an aspect of figured weaving. The sequence of paired images is not systematic, and the order of the samples is not necessarily from simple to complex. Samples are grouped by similarity of color so that the viewer is not distracted by "clashes." All textiles are shown at real size, with few exceptions that allow a motif or particular detail to be included.

Many of these textiles were woven centuries ago, when motifs tended to be reduced in size; the smaller scale allows multiple examples to be assembled on one page. Today's figured wovens, especially works of art, are larger and would not fit this limited format.

Every design device, structure, and technique pictured on these pages may be rethought with a contemporary aesthetic and employed to weave textiles that are relevant today, and can be woven on hand or industrial looms. The lessons these small samples contain are not historical: they are about surface, structure, patterning, and the multitude of opportunities the Jacquard medium offers.

Figure 1.7. Rose *by Anna Araldo, LFS.*

Figure 1.8. Facing page: Cosmetica *by Tuulia Lampinen, LFS.*

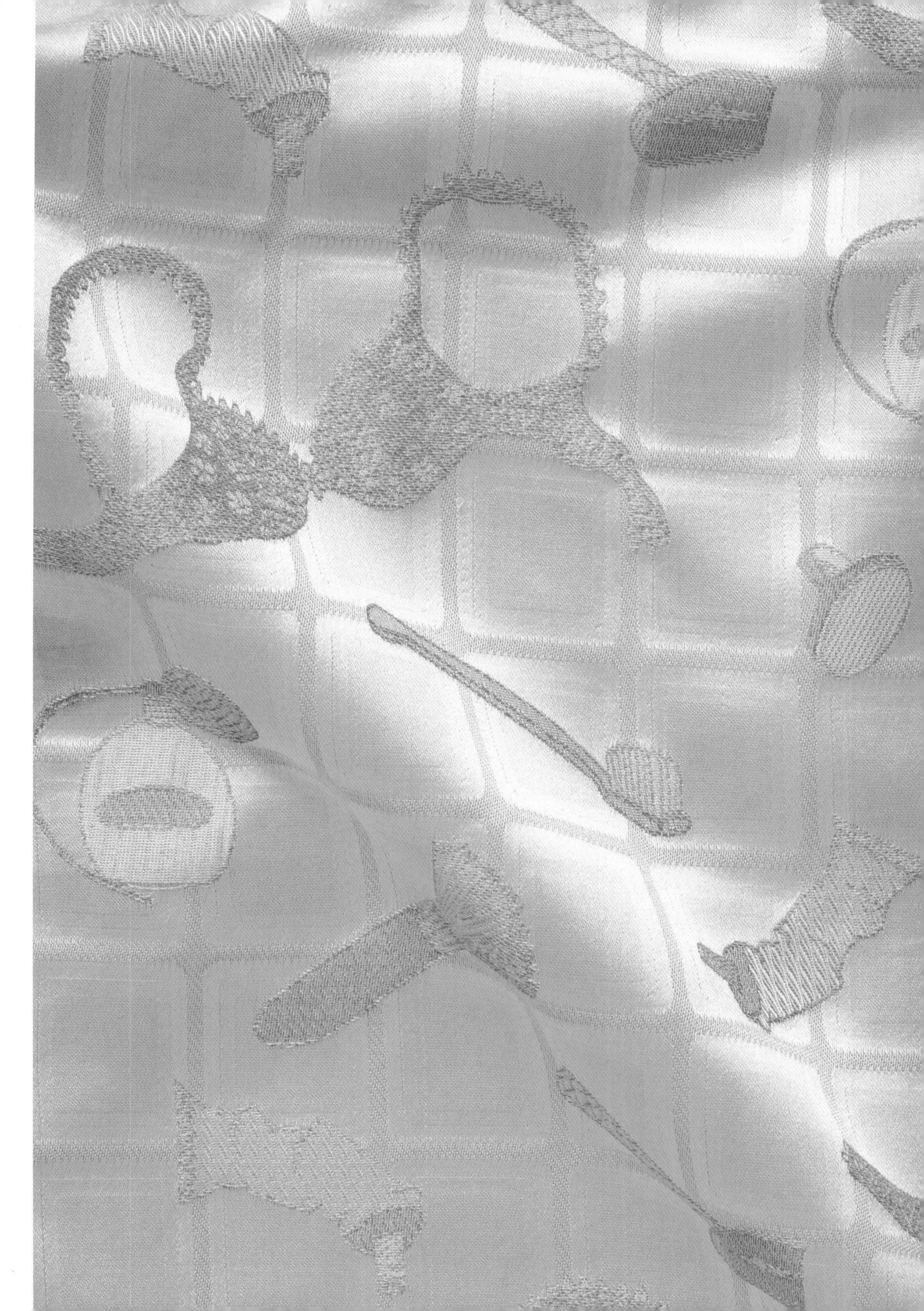

I. Motifs that rest above the ground

II. Textured motifs on plain or patterned grounds

III. Simple or complex patterning; simple or complex techniques with single or multiple warps and wefts series

IV. Textured and figured backgrounds beneath a patterned foreground

V. Expansion of space, brocading in gold and silver

VI. Perspective and a sense of depth (*following pages*)

Captions for textile example pages

I. Motifs that rest above the ground

Figure 1.9. *(pp. 10–11)* Row by row, from left to right: *Spitafields silk and silver brocade, Spitafields warp-patterned silk, both eighteenth century, Cora Ginsburg LLC;* 5 weft lampas, *Francesco Sala, LFS; eighteenth-century silk brocade, Societé Le Manach; eighteenth-century silk and silver* Bizarre *brocade, Cora Ginsburg LLC; seventeenth-century brocaded satin, CJH. Facing page: eighteenth-century brocaded Spitafields silk, Cora Ginsburg LLC.*

II. Textured motifs on plain or patterned grounds

Figure 1.10. *(pp. 12–13)* Left: Martine, *figured silk textile with three self-patterning wefts, Martine Peters, LFS;* right, from top to bottom: Onions, *damask, Hans Thomsson, LFS.* Swirls, *damask with two self-patterning wefts, Narin Panitchpakdi, LFS; figured trim, Laura Nicholson, CJH;* Lemon Leaves, *damask with two self-patterning wefts, Martha Porter, LFS.* Facing page: Copper, *damask with two self-patterning wefts, Narin Panitchpakdi, LFS; figured textile with warp patterning, self-patterning, and pattern wefts, Cora Ginsburg LLC.*

III. Simple or complex patterning; simple or complex techniques with single or multiple warps and wefts series

Figure 1.11. *(pp. 14–15)* Left to right: Grenade, *damask with two self-patterning and two pattern wefts, Emelia Haglund, LFS; eighteenth-century warp and weft-patterned textile, Societé Le Manach.* Facing page, from left to right, row by row: *eighteenth-century damask, Cora Ginsburg LLC;* Lines, *damask with two self-patterning wefts, Kazuyo Nomura, LFS; eighteenth-century Spitafields brocade with self-patterning wefts, Cora Ginsburg LLC;* Facett, *damask with three self-patterning wefts, Morgan Bajardi, LFS; eighteenth-century lampas, and brocaded* Bizarre *silk, Societé Le Manach.*

IV. Textured and figured backgrounds beneath a patterned foreground

Figure 1.12. *(pp. 16–17)* From left to right by rows, top to bottom: Small Squares, *damassé, Antoinette Stucky;* Tove's Velvet, *2-pile voided uncut velvet, Tove Andéer;* Bobles, *figured rib ground with three self-patterning wefts, Stine Øestergaard;* Martin's Drouget, *Martin Ciszuk, LFS. Eighteenth-century cannelé with pattern wefts, Cora Ginsburg LLC,* Pine Motif, *figured rib ground with two self-patterning and one pattern weft, Tuulia Lampinen, LFS.* Facing page: *eighteenth-century brocaded lampas with self-patterning wefts, Societé Le Manach.*

V. Expansion of space, brocading in gold and silver

Figure 1.13. *(p. 18) Silk and gold* Bizarre *textile, eighteenth century. Courtesy of Cora Ginsburg LLC.*

Figure 1.14. *(p. 19) Silk and silver Spitafields brocade, eighteenth century. Courtesy of Cora Ginsburg LLC.*

VI. Perspective and a sense of depth

Figure 1.15. *(pp. 20–21)* Left: *eighteenth-century brocaded tabby;* Right: *eighteenth-century silver and silk lampas. Courtesy of Societé Le Manach.*

VII. Complex patterning and technique

Figure 1.16. *(pp. 22–23) Eighteenth-century silk and silver brocade. Courtesy of Societé Le Manach.*

weft 2
weft 1
weft 2
weft 1
Pattern
Pattern
all down
broken twill
Weft 1: broken twill
Weft 2: all down

2 Sample analysis and documentation: structure and design

Learning to analyze an existing sample and record the data necessary for its reconstruction trains a designer in a number of important skills. Sample analysis hones the designer's ability to identify warp and weft series, recognize and draft weaves, and ultimately to gain a deeper understanding of how figured textiles are constructed.

There are two phases in sample analysis: first observation and note taking, then the compilation of a record sheet. What information is important to obtain and record? Structural information tends to be the most important data for makers, followed by material and design data. Details of provenance may be more useful for textile historians and conservators. Depending on the reader's objectives, analysis can be limited to the observation and drafting of one or two weaves of a given sample, or include material and design data.

Analysis process and records described in this chapter may be used by weavers and scholars, and should be adapted to individual needs and objectives. Steps for collecting and recording structural data are given in detail, followed by a brief summary of the contents of a design record.

Analysis of fiber content and yarn construction is not reviewed, as it is assumed that readers are already familiar with these topics.

Figure 2.1. Facing page: *Tools used for sample analysis. Figured textile with one warp and two self-patterning wefts.* Sticks *by Tuulia Lampinen, LFS.*

Observation and annotation of structural data

The following tools are needed for sample analysis: note and squared paper, ruler, pencil and eraser, teasing needle or long pin, and a pick glass. For very fine textiles, a basic digital camera with a macro lens function is useful. If the fabric is woven with multiple warp and/or weft series, colored pencils or pens are useful for drafting each series with a distinct color. Needle, thread, and pins may be used for delimiting weave repeats on the swatch.

Following the steps described, observe and note down data, then compile an organized record of the analysis. A comprehensive record sheet would include the data listed in Figure 2.3. If obtaining weave data is the sole objective of the analysis, examine and draft the weaves on a simplified record sheet, noting which weave is the ground effect and which is the patterning.

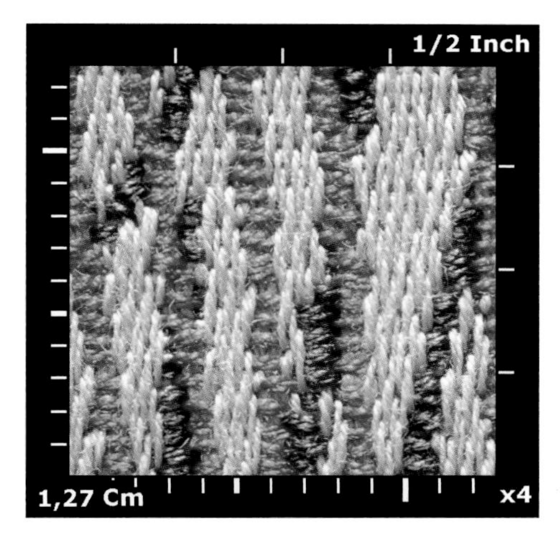

Figure 2.2. *Detail of* Sticks.

Title
- Date of sample analysis
- Provenance
- Date of manufacture
- Technique

Warp and weft
- Warp series, proportion and sequence, sett and materials
- Weft series, proportion and sequence, sett and materials

Weave data
- List and number of weaves effects: ground or pattern
- Verbal description, drawdown and/or sectional drafts of all weaves

Additional data (optional)
- Loom data
- Finishing or embellishment
- Historical, geographical, etc
- Other

Figure 2.3. *Comprehensive analysis record.*

Step 1: Cut out a swatch from the textile to be analyzed in an area that contains one or more repeats of each distinct weave structure.

Step 2: Determine which side of the cloth is the face; generally, this is the more attractive side with clearer patterning. If the textile is reversible, describe a distinguishing factor of the side on which the analysis will be made, such as: "the darker side is considered the face."

Step 3: Establish which series of threads is the warp and which is the weft. To identify the warp, determine which series of threads is stronger and better suited to support the stresses of weaving. The warp is the smoother of the two series, composed either of continuous filaments, or if spun, the fibers are usually longer than those of the weft; the yarn may be plied and more compact, so as not to fray or catch during weaving. The warp is the more elastic of the two series of threads, and when compared with the weft, the warp shows more crimp. Some samples, when held against a light source, show slight divisions or regular groupings between threads; these are called reed marks and indicate the warp. Fiber content is a factor for identifying warp and weft, as materials differ in strength and elasticity. The stronger and more elastic fiber is usually found in the warp direction. Once warp and weft direction is established, position the textile with the warp aligned vertically on the work surface.

Step 4: Record warp data.

1. Number of warps: determine if there are one or more warp series and the function of these: ground, inner, binding, complementary, pattern, pile, and so forth. (A series is a group of threads, warp or weft, having a distinct function. For a better understanding of warp function, see Chapter 7.)
2. Proportion and sequence: if there are multiple series, ascertain and note down the proportion between ends of each series, and sequence. Examples: 2 ground/1 binding/2 ground, or 4 ground/1 pile/4 ground, and so forth.
3. Warp sett: to calculate warp sett, place ruler perpendicular to warp, then count the number of ends in one centimeter or inch. Use a pick glass and teasing needle if the textile is very fine. The setts of additional warp series are based on the ratio of each series to the ground warp. Example: if a textile has 4 ground/1 binding warp at forty ground ends per centimeter, there are ten binding ends per centimeter.

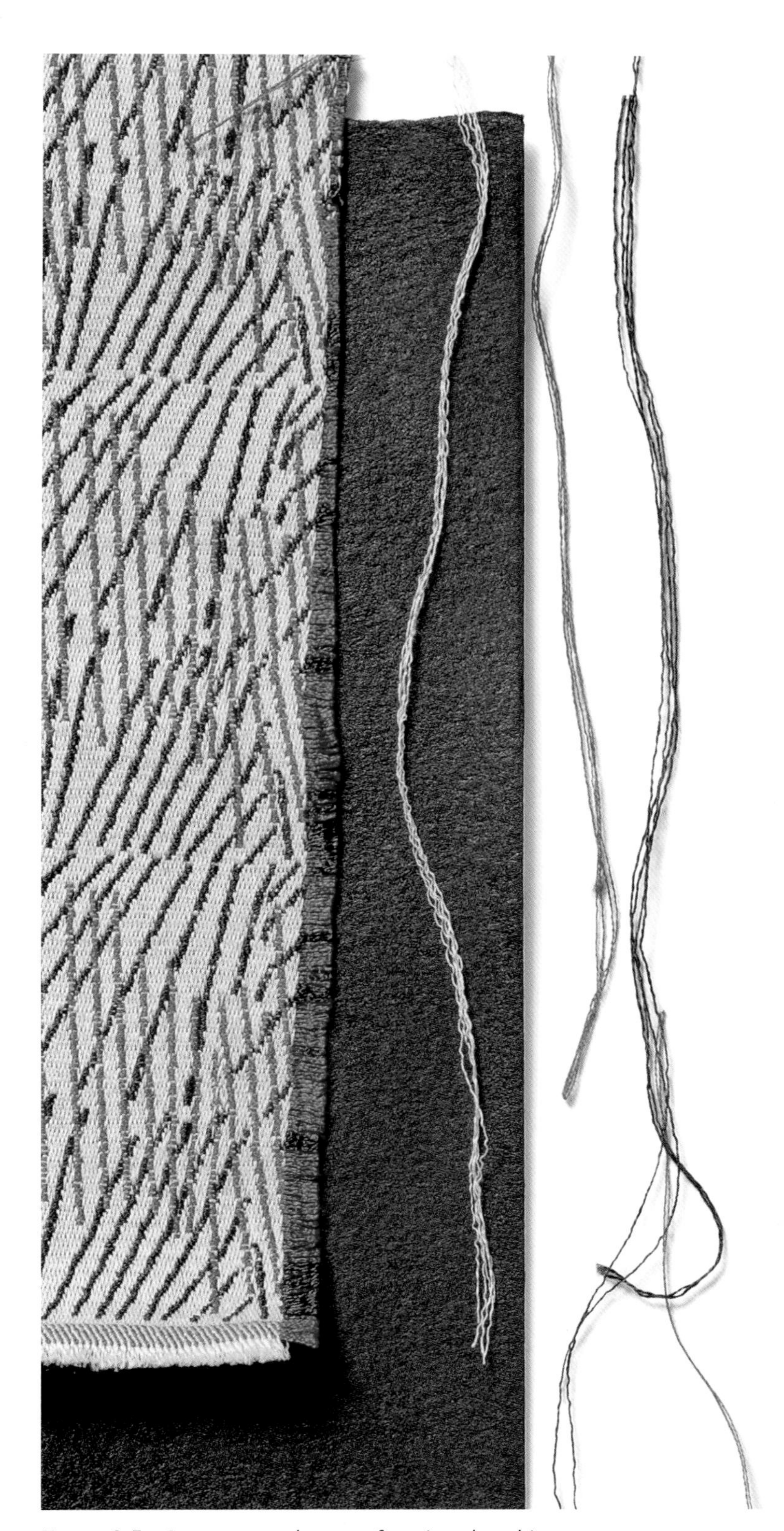

Figure 2.5. *One warp and two weft series; the white warp shows more crimp.*

Figure 2.4. Facing page: *Face and reverse of* Blue Boxes *by Elizabeth Tritthart, LFS.*

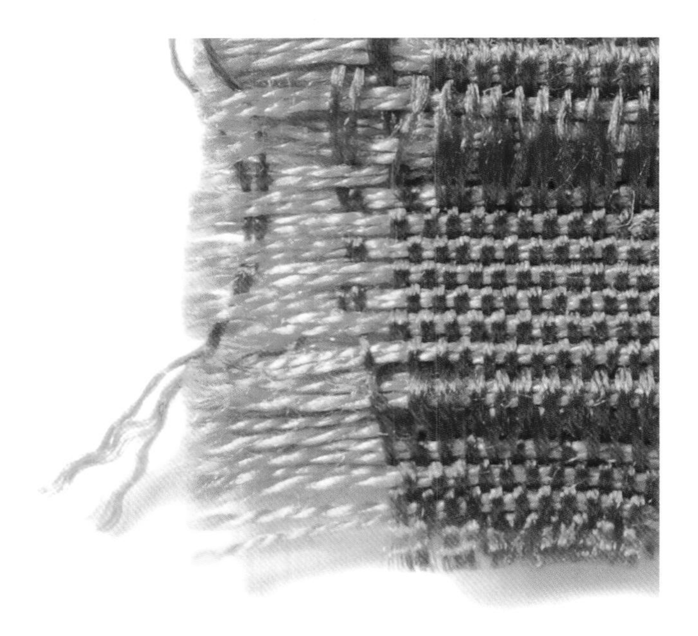

Figure 2.6. *Figured plain weave derivative with warp floats, CJH.*

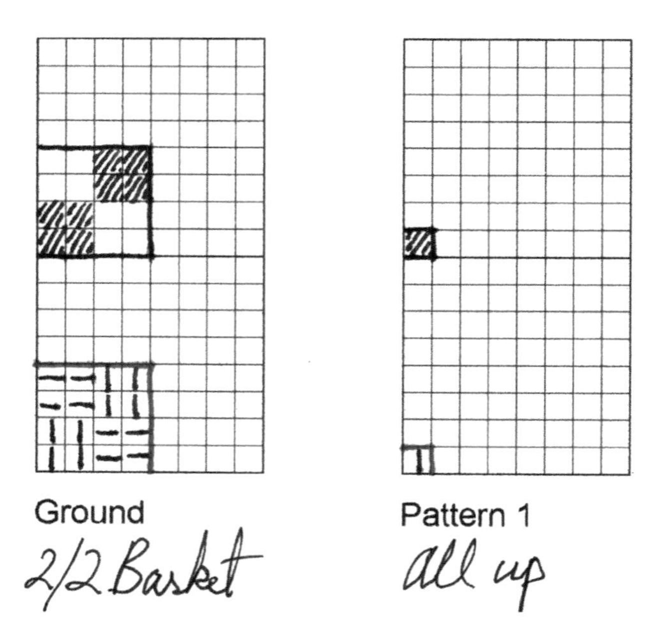

Figure 2.7. *A vertical mark indicates end over pick; a horizontal mark records end below pick. A drawdown is drafted using filled squares to represent raised warp ends; a missed square represents a warp end below a pick.*

4. Materials: examine warp materials and note: fiber content, yarn construction, and color. If a warp design is present, such as a stripe or chiné effect, record this data.

Step 5: Record weft data.

1. Number of wefts: determine the number of weft series. If more than one, ascertain function: ground, complementary, pattern, and so forth (to better understand definitions of function, see Chapter 7).
2. Proportion and sequence: if there are multiple series, note down proportion and sequence. Example: 1 ground/1 pattern weft, or 2 ground/1 pattern weft.
3. Weft sett: align ruler perpendicular to the weft, then count and record the number of ground picks in one centimeter or inch, using a pick glass and teasing needle if needed. The setts of additional weft series are based on the ratio of the ground weft to each series. Example: if the ground/pattern weft ratio is 1:1 at forty ground picks per centimeter, the pattern weft sett is forty, whereas with a ratio of 2:1 between ground and pattern wefts at forty picks per centimeter in the ground, pattern weft sett is twenty per centimeter.
4. Materials: examine weft yarn(s) and note fiber content, yarn construction, and color of each distinct yarn composing the weft. If a weft design is present, record this data.

Step 6: Identify each distinct weave structure (also called a weave effect, or simply "effect"), distinguishing between ground and pattern effects. Note: a ground effect is the weave area perceived as the background of the design, and is not necessarily the ground weave. See Chapter 7: ground weave, ground effect.

Draft the ground effect's weave structure on squared paper. If possible, find an area at the edge of the swatch, then remove several threads and picks to expose the ends of both. Using a teasing needle or long pin, carefully shift one warp end away from the weave, being careful to leave it interlaced with the weft. Place the pick glass over the loosened thread to obtain a clear view of the end's position. On a column of squared paper, one square and one point of interlacement at a time, plot the position of the warp end relative to each pick in the weft repeat. Use a vertical line or filled square to indicate warp over weft, and a horizontal line or blank square to indicate the warp under the weft. Adjusting the pick glass as needed, continue to plot the interlacement of each successive warp end until one complete repeat of the interlacement cycle or weave has been recorded. Delimit the weave repeat with a pen or pencil. If vertical and horizontal lines have been used to plot the interlacement order, convert this information into a drawdown on squared paper, using a filled square to represent a raised warp end, leaving empty squares to represent a warp under weft pick. Pick out and draft the remaining weave effects.

Note: If warp or weft setts are high and yarns fine, a macro photograph can be used to visualize and plot the interlacement of warp and weft.

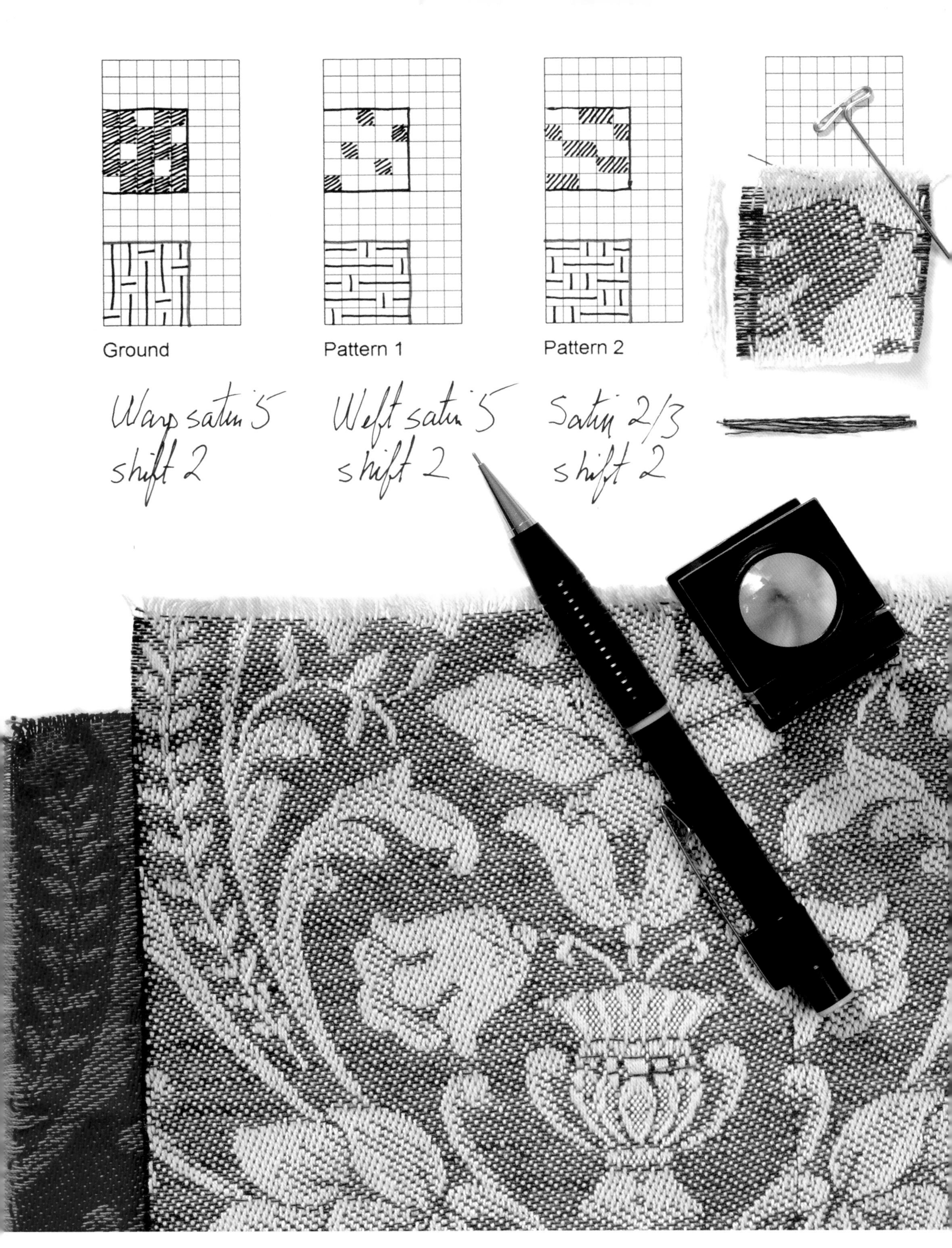

Figure 2.8. *Reversible linen damask woven with 5-end satins, CJH.*

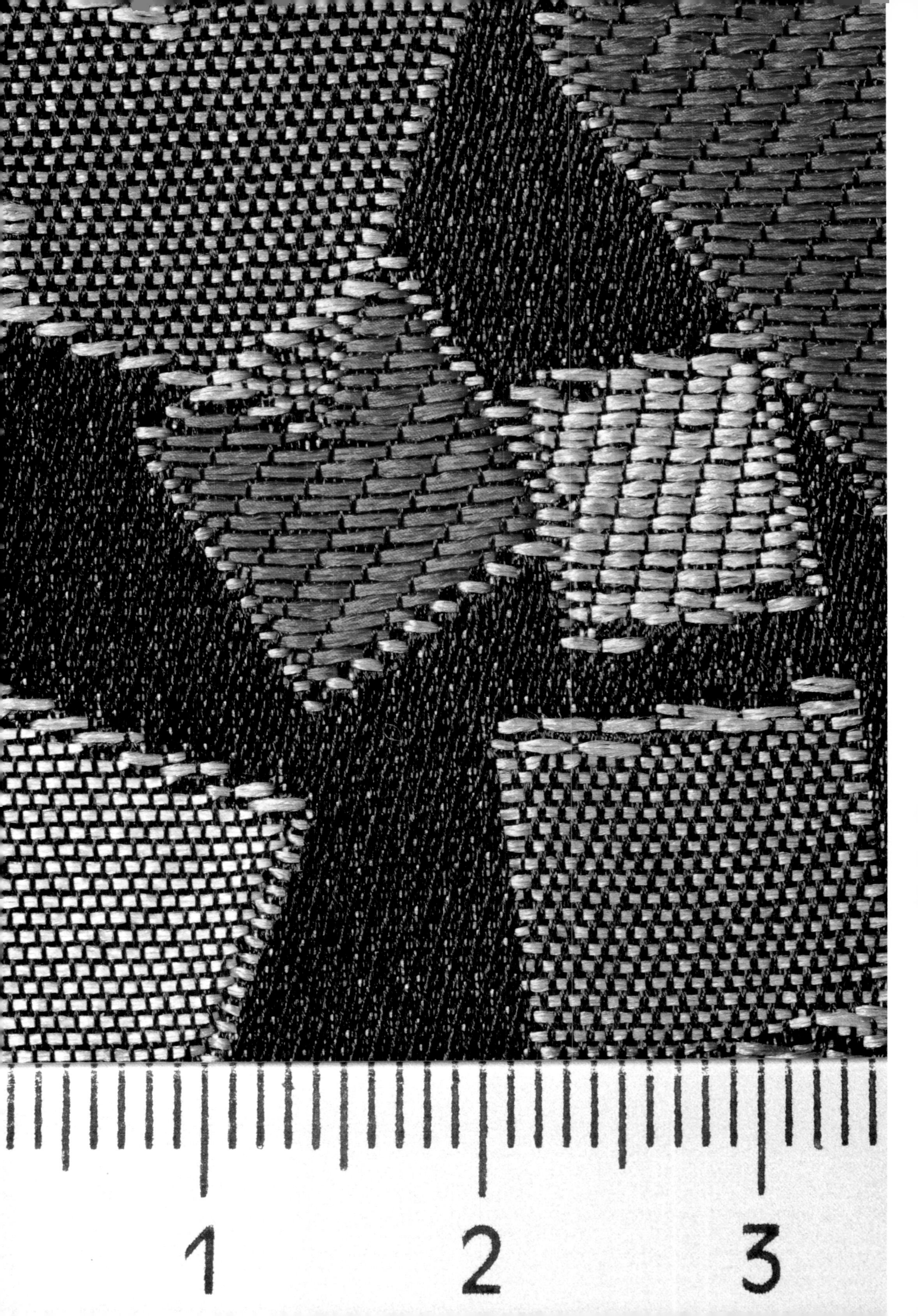
1
2
3

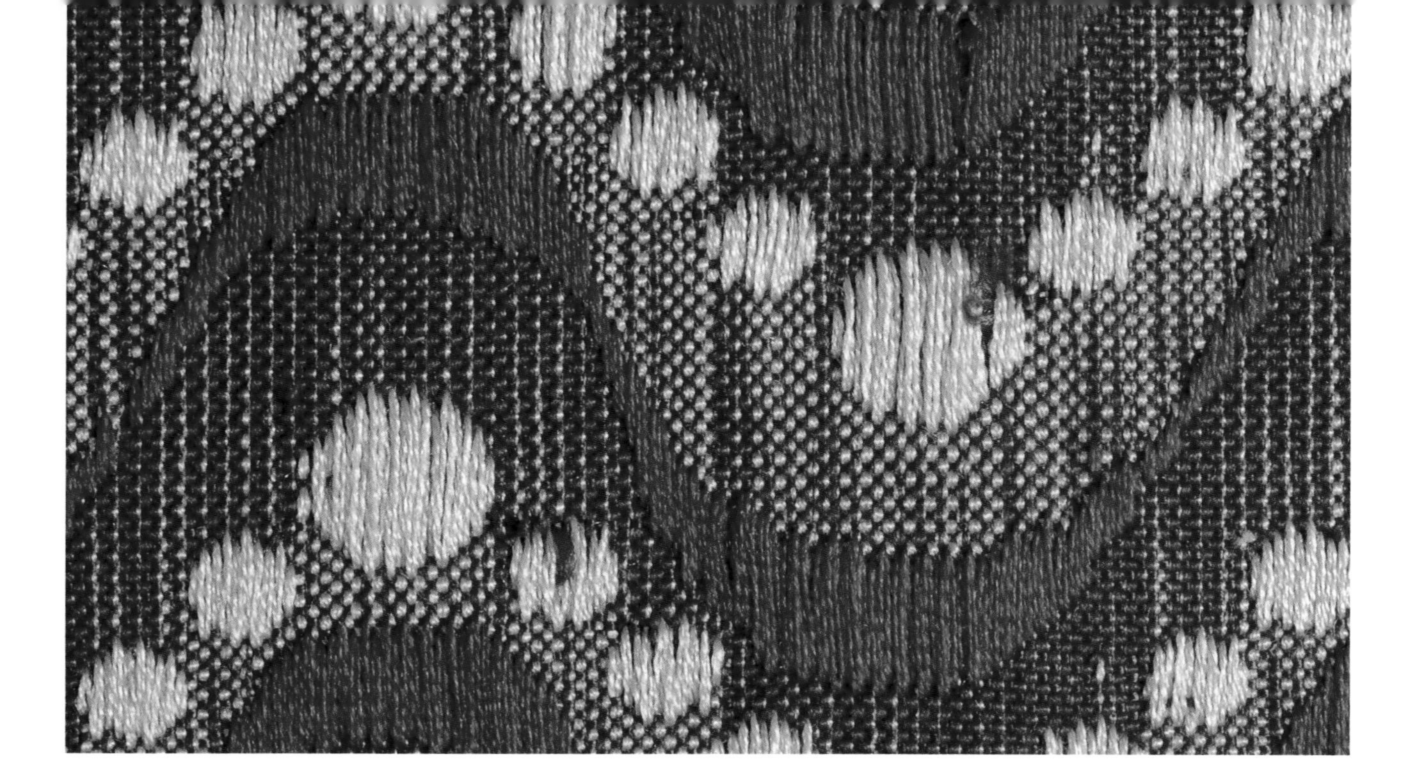

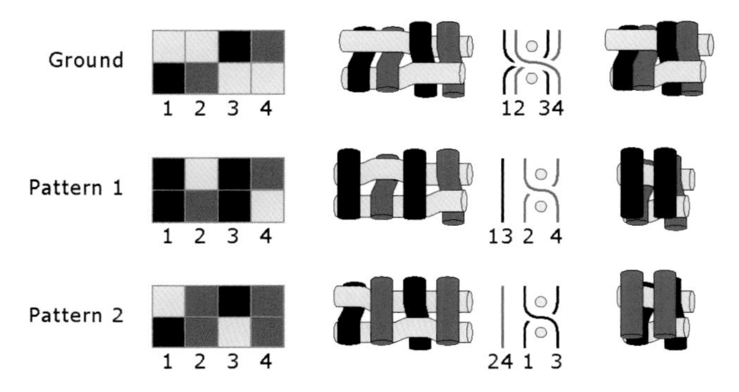

Figure 2.10. *The first warp series is drafted using a black pen; the second warp series is drafted in red. Weaves, detail, and photograph at real size for* Haitienne *by Martin Ciszuk, LFS.*

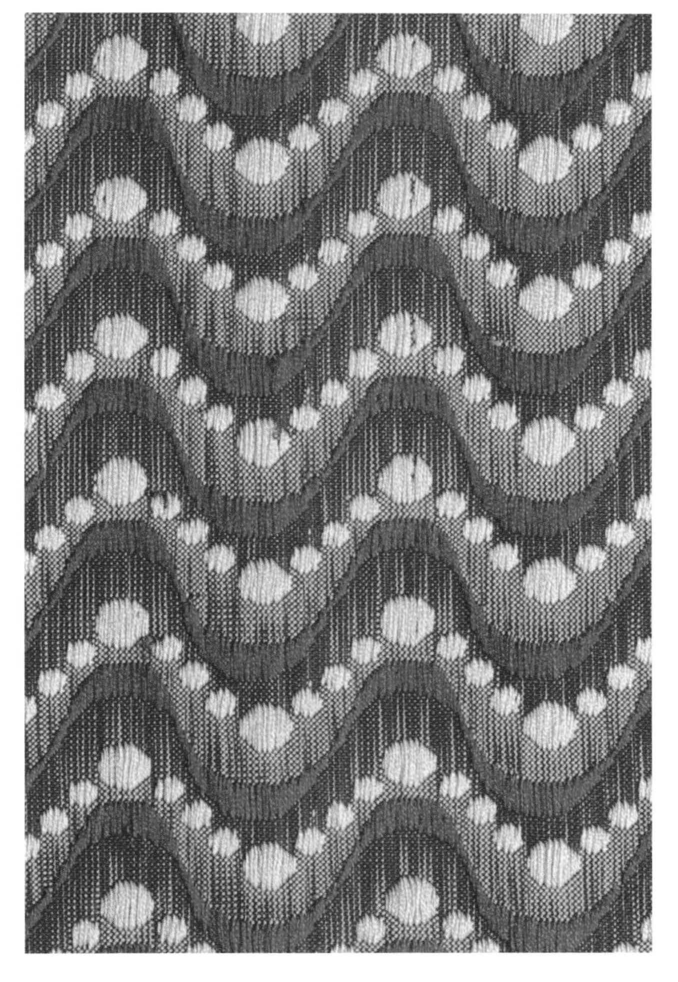

Figure 2.9. Facing page: *Macro photograph, useful for sample analysis of textiles with high warp and weft setts.* Blue Boxes *by Elizabeth Tritthart, LFS.*

If the textile is woven using multiple warp series, draft each warp with a distinct and preferably dark or saturated color, using a vertical line or filled square to indicate warp over weft. Use lighter colors, one for each weft series; draft warp under weft with a horizontal line, or an empty to light-colored square.

Though less used today than weave drawdowns, with practice, sectional drafting (explained in Chapter 3) is simple to use and easy to read when recording compound weaves. Note: a weft section shows the warp in profile and the weft threads "in section." A warp section shows the weft in profile.

Step 7: If compound weave structures are found in the sample, compile a reading note or table showing the weave structure corresponding to each weave effect, decomposed into the simple weaves used for each warp and weft series as shown in Figure 2.11. Weaves may be described verbally and/or graphically.

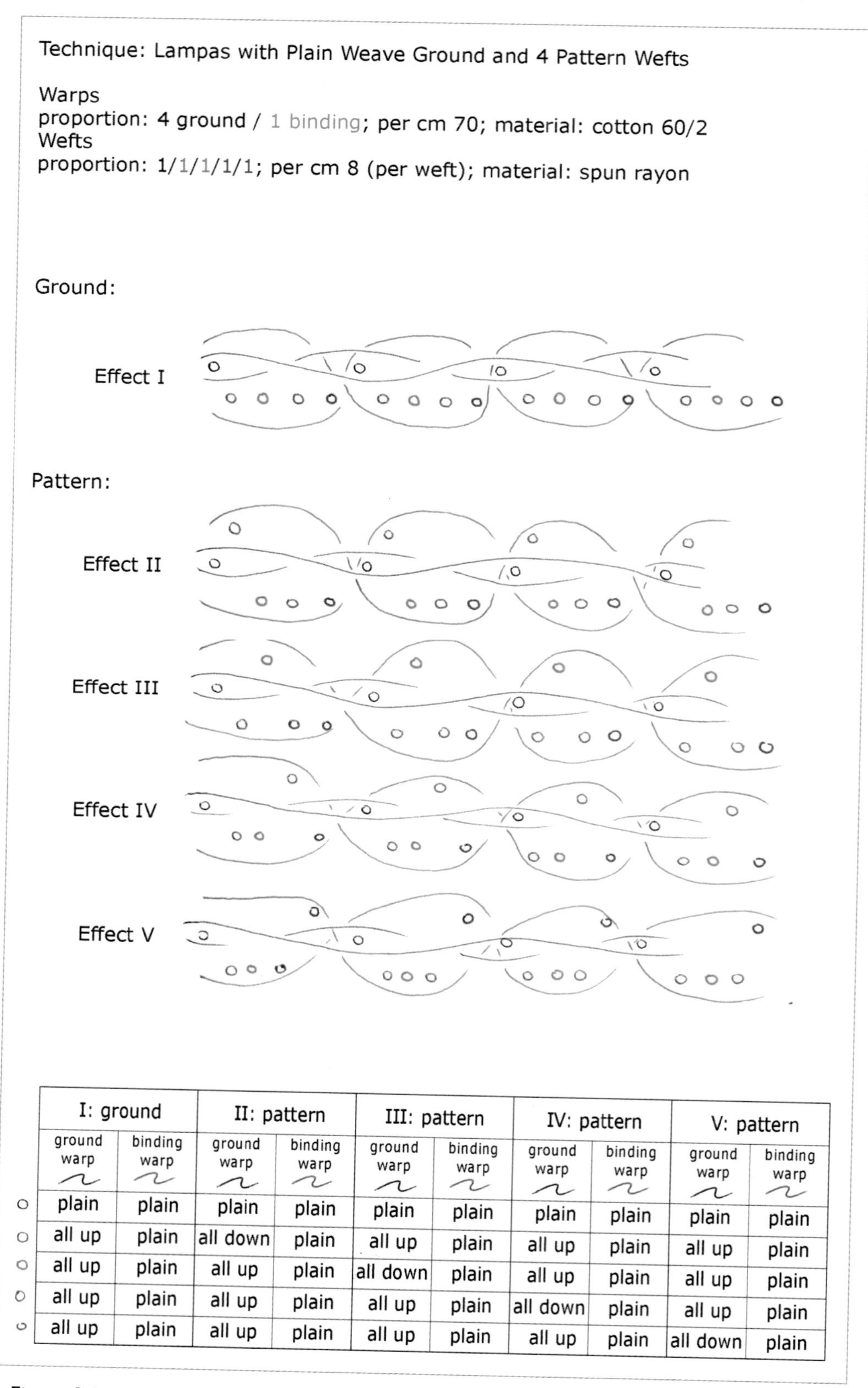

	I: ground		II: pattern		III: pattern		IV: pattern		V: pattern	
	ground warp	binding warp	ground warp	binding warp	ground warp	binding warp	ground warp	binding warp	ground warp	binding warp
o	plain	plain	plain	plain	plain	plain	plain	plain	plain	plain
o	all up	plain	all down	plain	all up	plain	all up	plain	all up	plain
o	all up	plain	all up	plain	all down	plain	all up	plain	all up	plain
o	all up	plain	all up	plain	all up	plain	all down	plain	all up	plain
o	all up	plain	all up	plain	all up	plain	all up	plain	all down	plain

Figure 2.11. *Weft sections and reading note for a lampas with two warp series and five wefts.*

Figure 2.12. 5 weft lampas *by Francesco Sala, LFS.*

Step 8: Note other details of textile, pertinent to its construction, such as the type of loom, if available or possible to hypothesize based on collected data. Finishing, embellishment, historical data, and so forth are also noted.

Step 9: Transcribe the collected data to a paper or digital record sheet. Include swatch, photograph, or digital image of textile with warp aligned vertically on the page.

Design analysis and documentation

Recording precise design data may be of secondary importance to many, and can require an in-depth understanding of figuring technology and techniques. If terminology used for recording design data is unfamiliar, review the glossary and discussion of design process found in Chapter 5.

Before collecting design data, complete and have available a structural record sheet for the figured textile to be analyzed. Obtain a full repeat of the design, if possible. Have available a measuring tape, pick glass, notepaper, and pencil. A basic camera or drawing materials are used to record the design.

The data listed in Figure 2.13 comprise the design record and may be modified to meet each user's needs and objectives.

Step 1: Photograph or draw the design. Delimit one complete unit on a printout, photograph, or drawing.
Step 2: Measure and record the motif's width and height. If the motif repeats, describe or sketch the repeat mode. Example: straight repeat, horizontal point repeat, half-drop, and so forth.
Step 3: Carefully observe the weave data on the structural record sheet and examine the areas of exchange between weave effects to obtain data relative to warp and weft. If the textile is woven with a single warp and weft series using an industrial loom, and the exchange between weaves occurs in units of one end, one pick, then the warp line sett is equal to the ends per centimeter, warp line value is 1, and step is 1. The weft line sett is equal to the picks per centimeter, weft line value is 1, and step is 1. If the exchange between weave effects occurs at intervals of more than one end or pick, ascertain the minimum interval, or unit of exchange in both the warp and weft direction. This minimum unit of exchange, or pattern unit, is the warp or weft step. See the explanation of *step* in Chapter 5.

If the textile is woven with more than one warp and weft series, and/or woven with older figuring devices or a combination of ground and patterning harnesses, warp and weft line sett will not be equal to the warp and weft sett data recorded on the structural analysis sheet. Warp and/or weft line value will no longer be equal to 1, and step may be more than 1 in either the warp or weft direction, or both. See Chapter 5 for explanation of warp lines, warp line sett, value, and step; and weft line sett, value, and step.

Design data
Motif:
- Repeat: width and height
- Repeat mode

Warp:
- Total no. warp lines
- Warp line sett and value
- Pattern step

Weft:
- Total no. weft lines
- Weft line sett and value
- Pattern step

Loom data:
- Figuring device

Figure 2.13 *Data relative to a figured textile's design.*

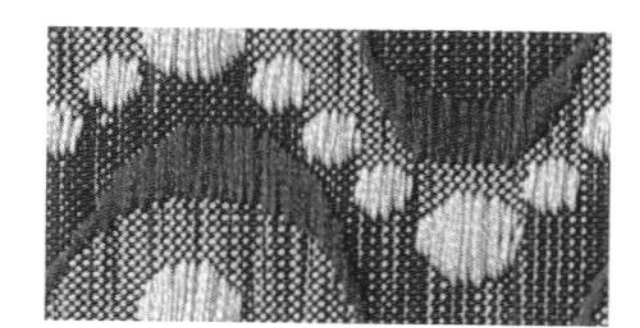

Figure 2.14. *One repeat of the motif of* Haitienne; *see analysis record in Figure 2.16.*

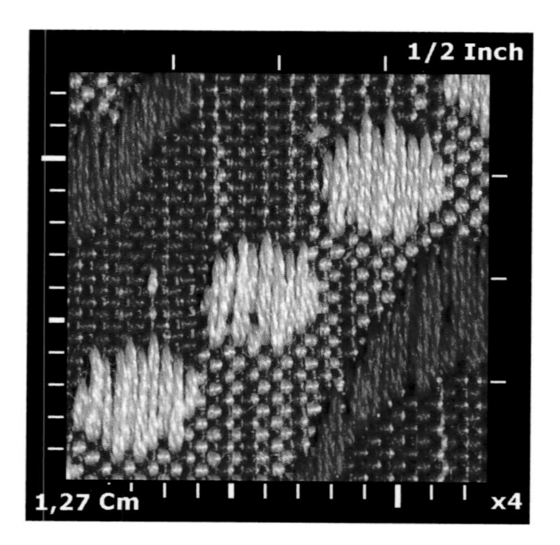

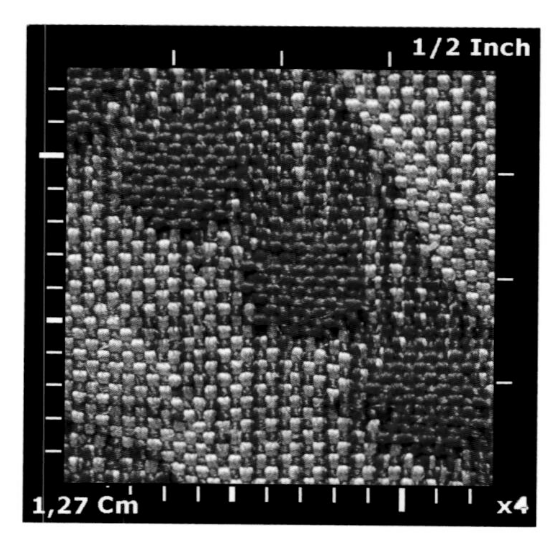

Figure 2.15. *Details of face and reverse of* Haitienne, *useful for establishing warp and weft data.*

"Haitienne"
Broderie fil à fil

1 Warp in two colors; proportion 1/1;
material: shappe 200/2; 54 ends/cm
1 Weft; material: silk; 20 wefts/cm step: 1

effect 1: ground, Louisine by warp and weft

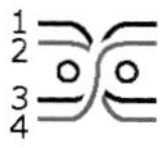

effect 2: floats of even ends
uneven ends and weft make tabby

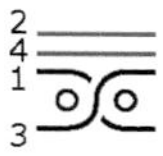

effect 3: floats of uneven ends
even ends and weft make tabby

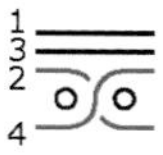

Design repeat: 3,4 x 2 cm 96 cords
50 passes

Loom setup: Jacquard and split harness with 16 bannisters
controlled by dobby head
4 straight repeats of 384 hooks with single cords

Reading note: weaving face up
1 cord of graph paper = 2 hooks
1 pass of graph paper = 1 card

	I: ground		II: pattern		III: pattern	
	1, 3	2, 4	1, 3	2, 4	1, 3	2, 4
o	plain	plain	plain	up	plain	up

Louisine on bannisters

Figure 2.16. *Structural and design data for* Haitienne, *a figured 2/1 basket (traditionally called a* louisine*), with two self-patterning warp effects.*

Step 4: Optional: note known or hypothetical data relative to the figuring device used to produce the design, such as the total number of hooks or figuring ends, harness tie, etc.

Haitienne was woven on an older, card-driven hand Jacquard loom, mounted with a split harness, that is, with both a harness controlled by a dobby machine to produce the ground effect, and a second, figuring harness to produce the patterning.

Note: Often the record sheets in this book employ the term *cord* in place of warp line and *pass* in lieu of weft line (see Chapter 5).

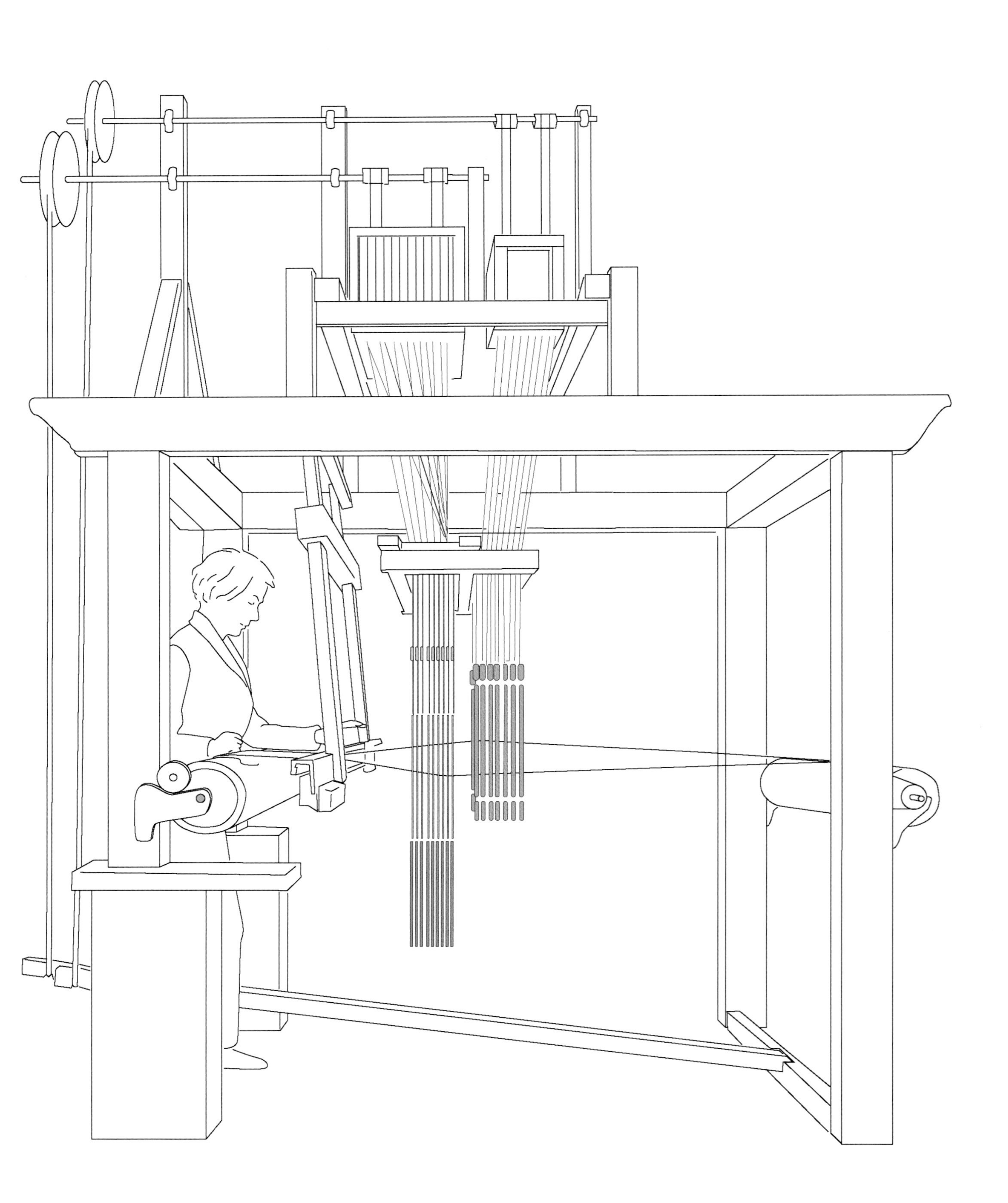

3 Weave drafting methods: recording, designing, and reading weaves

Weave drafting serves a designer for recording existing structures and constructing and sampling new weaves for figured textiles. Knowledge of the conventions of weave representation and the relationship between representation and the physical weave opens the door to textile literature and archives, past and present.

This chapter reviews the objectives of weave representation, defines differences between illustration and draft, and demonstrates two widely used drafting methods for simple and compound weave structures. Several record sheets are included as examples of how the multiple weave drafts of a figured textile may be recorded; suggestions for converting older drafts to a more legible format conclude the chapter.

Relationship between weave representation and the physical textile: warp and weft alignment

What is the correlation between a woven textile and its representations on screen and paper? Whether on screen or printed page, the position of warp and weft and their interlacement is most often shown from the weaver's point of view when standing above the textile at the front of the loom. The bottom edge of the page or screen corresponds to this position. For reasons of legibility and space, sectional drafts may sometimes show the warp aligned horizontally on the page or screen, with the first weft to the reader's left.

Figures 3.2 and 3.3 illustrate alignment and interlacement of warp and weft. The perspective in all these illustrations is the same; the weave is viewed from above the textile and at the front of the loom. The warp is perpendicular to the breast beam; the first warp end is shown on the weaver's left. The weft is parallel to the beater and the first pick of the weave rests closest to the breast beam of the loom.

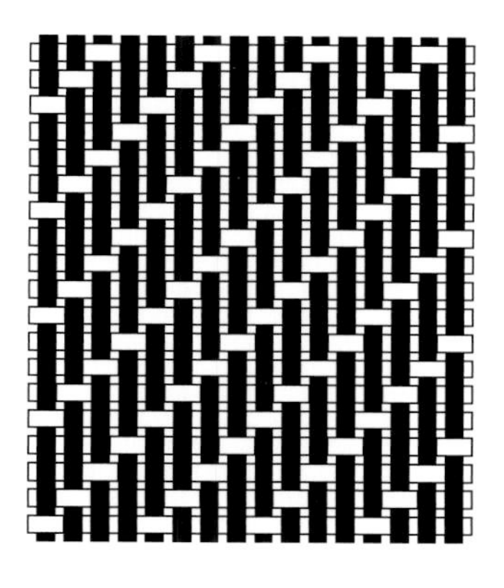

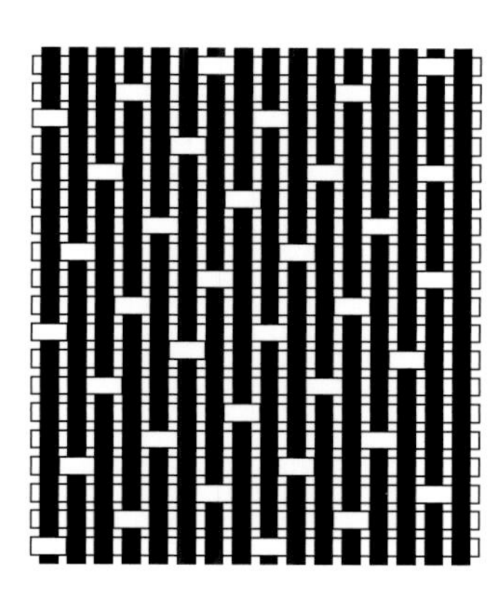

Figure 3.1. Facing page: *Weaver at front of loom. The warp is perpendicular to the breast beam, the first warp end is on the weaver's left; the first weft insertion rests both parallel and closest to the breast beam.*

Figure 3.2. *Illustrations showing a weave from each of the three weave families.* From top to bottom: *plain weave and derivatives, twills, and satins.*

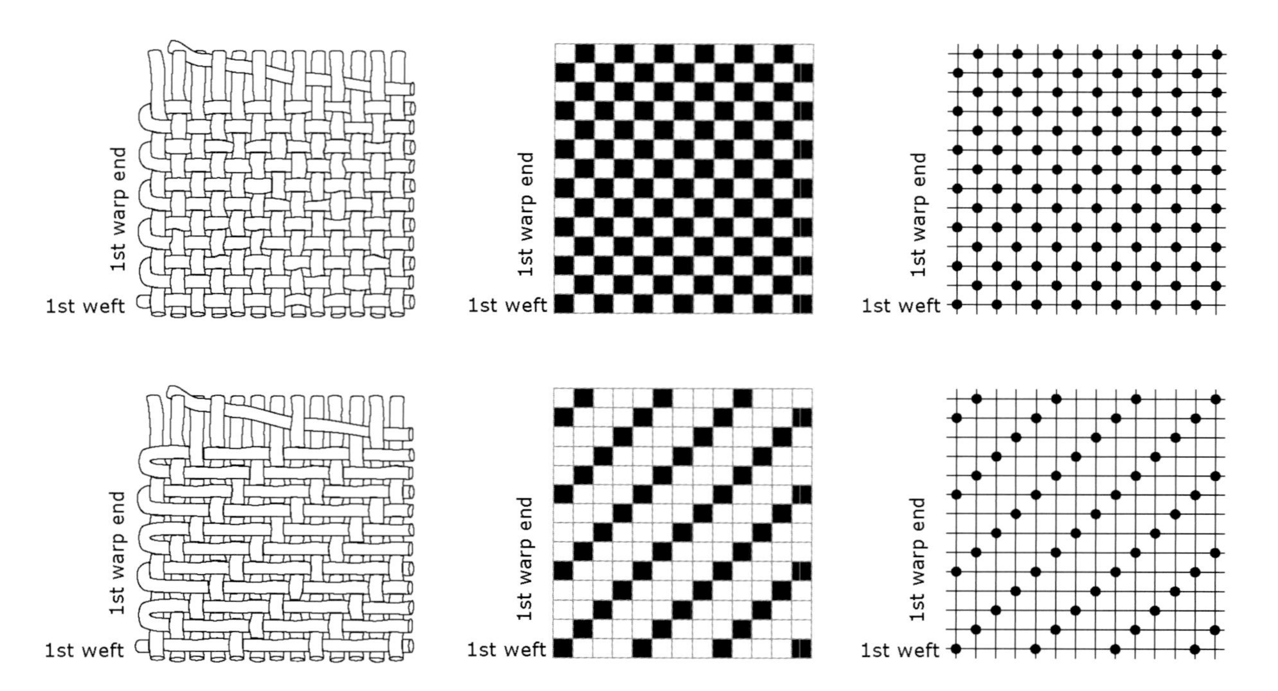

Figure 3.3. *Three of the many methods used for illustrating woven textiles are shown for plain weave and a 1/3 Z twill.*

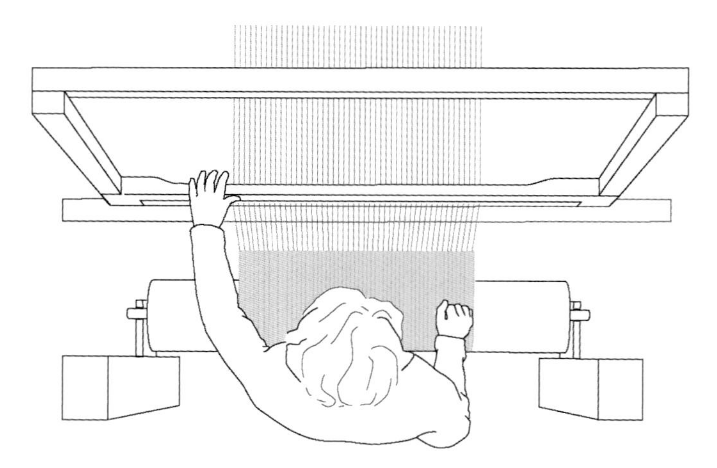

Figure 3.4. *Warp, weft, and weave as seen by the weaver at the front of the loom.*

As a textile is woven, at each point of intersection between warp and weft, warp ends rest above or below the single weft insertions, or picks. In each horizontal row of drawings in Figure 3.3, the warp's position over or under the weft is represented in three distinct ways.

1. The drawings on the left illustrate the appearance of the weave; warp and weft threads and their interlacement are represented in a realistic way. Warp position, over or under the weft, is not emphasized.

2. In the central group of drawings, the warp's position relative to weft is drafted on squared paper. Each vertical column of squares represents one end; each horizontal row of squares one pick. A filled square indicates warp over weft (warp up); white indicates warp under weft (warp down). Note that only the warp's action is marked with a sign—a filled square.

3. The drawings on the right follow yet another method of illustration less used today: vertical lines represent the warp, horizontal lines the weft. Each intersection represents one point of interlacement between warp and weft; marked points of interlacement indicate warp over weft; unmarked intersections show warp under weft. As in the central column of drawings, only warp action is indicated.

In all three representations, the warp is aligned vertically, the weft horizontally; the first weft is located at the lower edge of each figure, the first warp end is on the left. Note that the weave repeat—the minimum unit of interlacement order of ends and picks—is not delimited in any of the figures above. None of these illustrations is a weave draft.

Illustrations

Drawings of weaves, as well as simulations generated by a computer, help the designer to evaluate the appearance of a structure before sampling, or the overall effectiveness of combinations of weave and design. Photographs are employed to document the appearance of a weave or to compare weave effects, while macro photographs aid in weave analysis. All are useful for illustrating weaves, but none fulfill the very specific role of a weave draft.

Figure 3.5 *Drawings, simulations, and photographs of* Cypress Trees *by Sheetal Khanna-Ravich, LFS. Damask with two self-patterning wefts.*

Weave drafts

A draft is a technical drawing with a distinct function: it records a precise unit of information relative to one or more components of a textile. When designing weave-patterned textiles, the designer drafts a pointpaper, warp and weft sequences, as well as weaves. Yet other drafts or notes record warp and weft setts, color sequence, selvages, and so forth; together these complete the Jacquard file in preparation for sampling and production.

A **weave draft** records exact and complete information relative to the interlacement cycle of every warp and weft in a given weave repeat; this information, in combination with all the other drafts in a Jacquard file, is used to generate the program, or card that controls the warp's action during weaving.

Throughout this book, weaves are drafted as drawdowns and weft or warp sections.

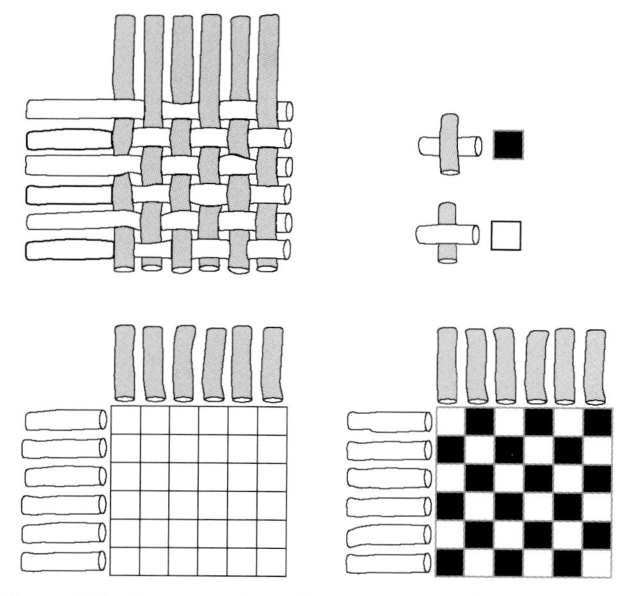

Figure 3.7. *Correspondence between squared paper, warp, weft, and warp position.*

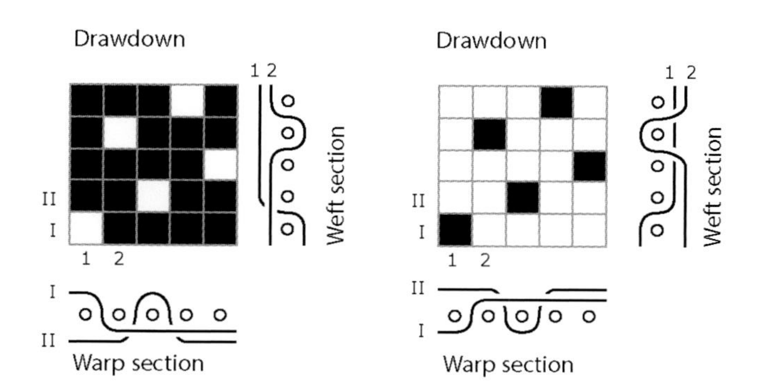

Figure 3.6. Facing page: *Warp and weft satins 5, shift 3 are used to weave the damask shown in the photograph.* Above: *Both weaves are drafted using drawdowns, weft section/ warp profiles, and warp section/weft profiles. Coral-colored damask. Courtesy of Cora Ginsburg LLC.*

Drafting simple weaves: drawdown and sectional draft

Drawdown

The most widely used weave draft is the drawdown, in which one complete weave repeat is drawn on squared paper. Each vertical column of squares represents one end; each horizontal row, one pick. Each square of a drawdown represents one point of interlacement between warp and weft. A filled square marks warp over weft; a white or empty square indicates warp under weft. In this book, a black or brightly colored square is used to represent warp up; a white or light-colored square indicates warp down.

The order in which raised ends/black squares alternate with lowered ends/light squares determines the weave structure. The minimum number of ends and picks necessary to complete one cycle of interlacement is the weave repeat. By convention, if more than one repeat of a weave is drafted, the minimum unit is indicated using a graphic device such as an outline, highlighting, or numbering the ends and picks of one repeat.

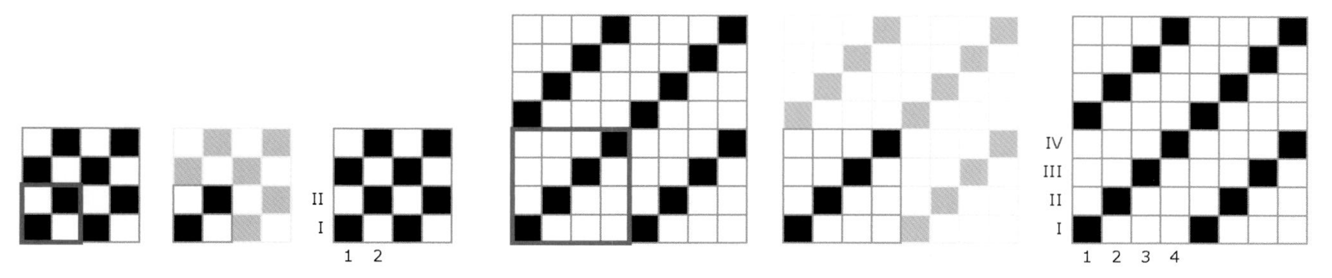

Figure 3.8. *One repeat of plain weave and 1/3 Z twill outlined, highlighted, and/or numbered.*

Sectional Drafting

Sectional drafting is particularly effective for describing the interlacement order of warp and weft and offers the advantage of a more dimensional visualization of warp and weft positions. With practice, sectional drafts can be easier to read and faster to draw than the more abstract drawdown.

Weft sections show the warp in profile, while warp sections show the weft in profile.

Throughout this book, weft sections are used in preference to warp sections, as it is the warp's action rather than the weft's that is controlled by the loom's patterning device, be it digital interface, hooks and cards, cords and pulleys, and so forth.

Figure 3.9. Top: *Drawdown with weft and warp sections of weaves used by Päivi Fernström for* Pomegranate, *a damask with three effects, LFS.* Bottom: *Photograph of* Pomegranate.

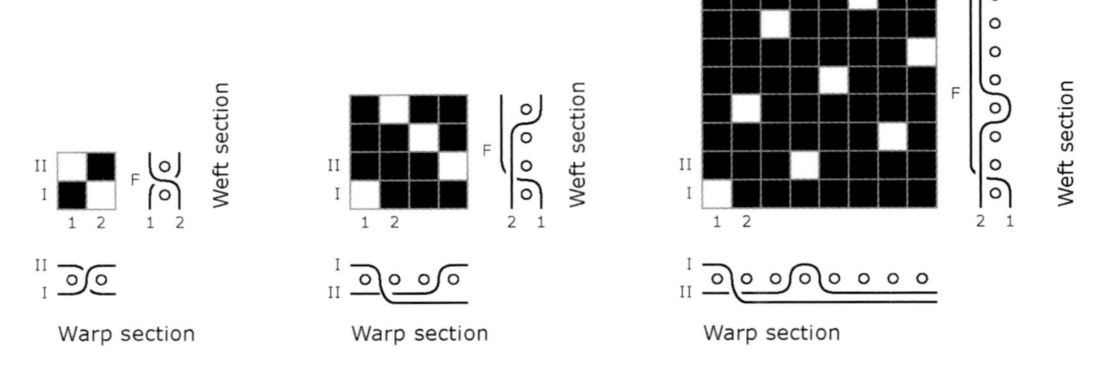

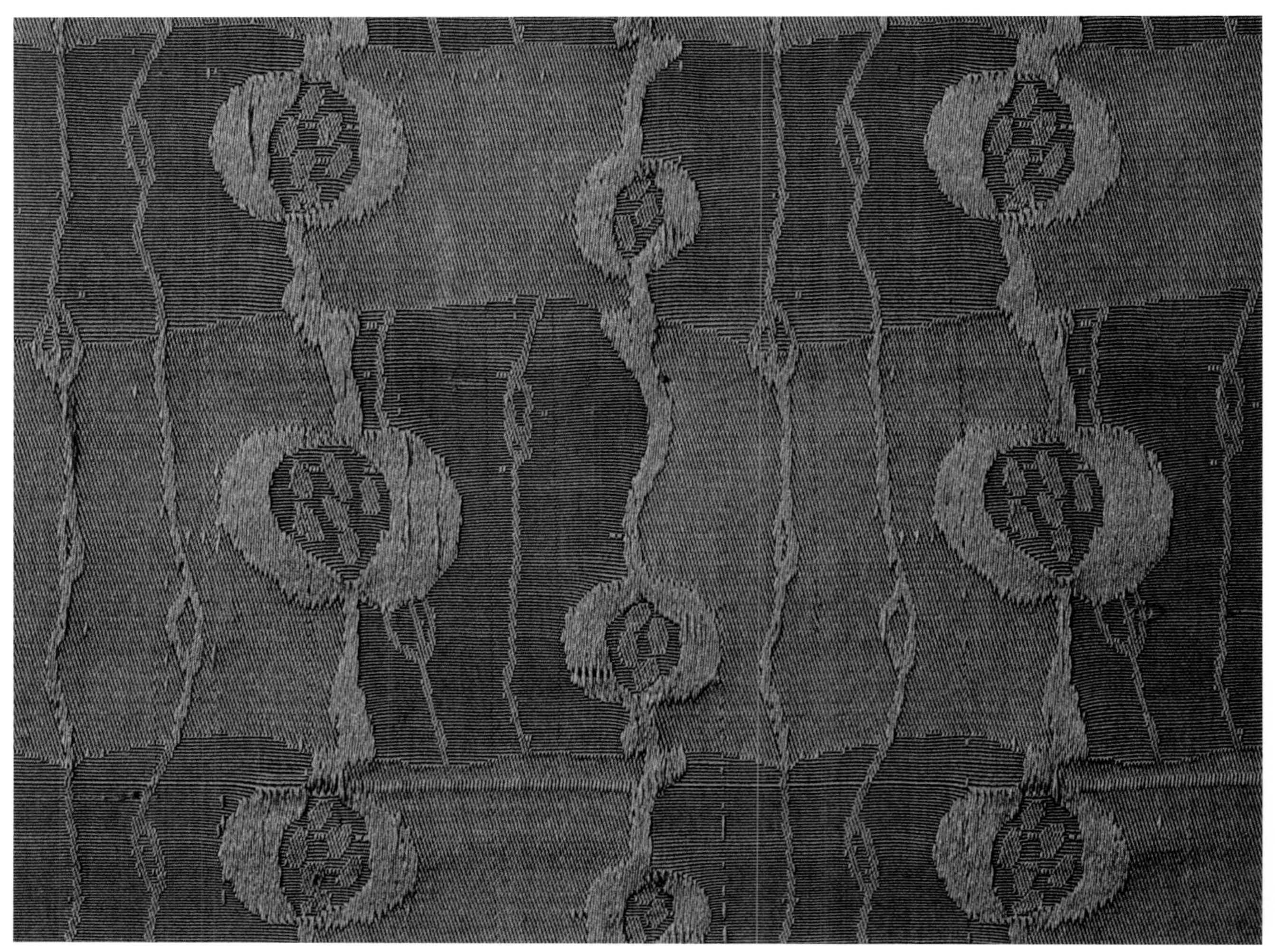

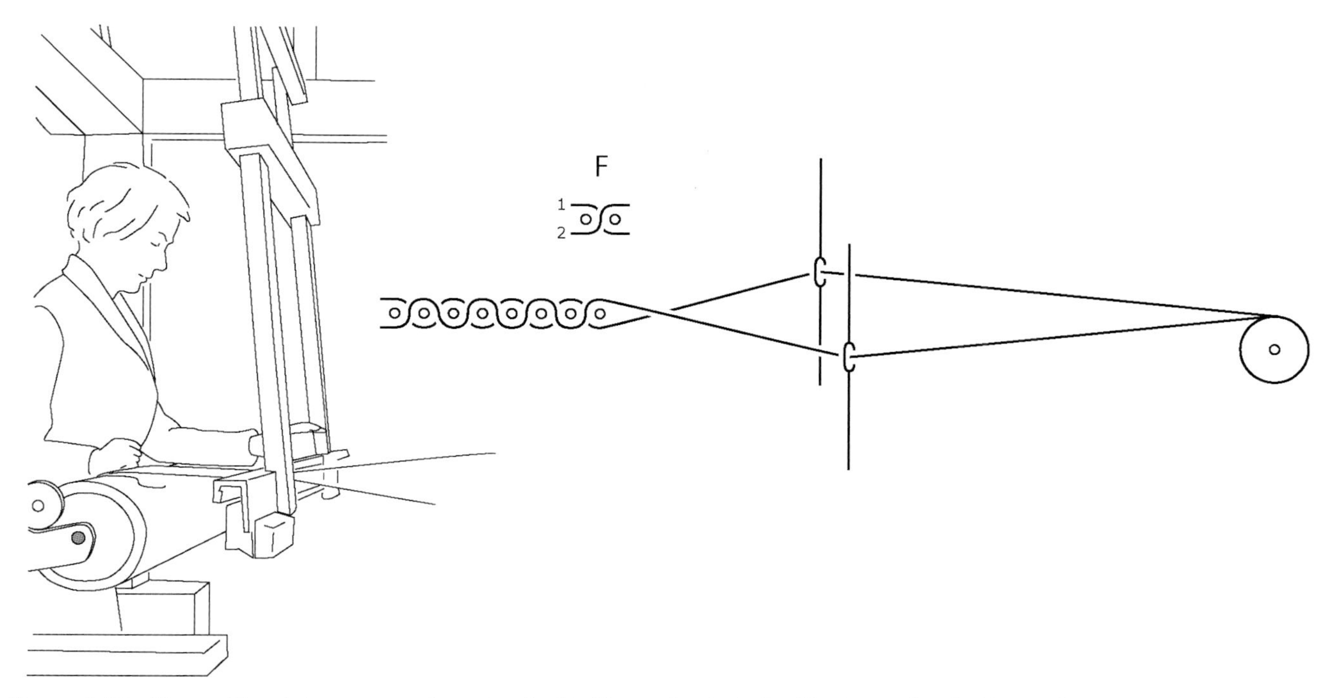

Figure 3.10. Above: *The plain weave weft section in this illustration is drafted horizontally; the first weft is nearest to the breast beam; the first warp end is to the weaver's left.*

When drafting weft sections of simple weaves, two ends (or two groups of adjacent ends that follow an identical interlacement order) and all picks of the weave repeat are sufficient for documenting most regular weave structures.

As with other forms of weave representation, it is useful to understand the correlation between the actual weave structure and sectional drafts on paper and screen. When a weft section is drafted, the warp is drawn in profile, as if seen from the right side of the loom. From this point of view, the first warp end lies on the left edge of the warp, farthest from the viewer; the first weft lies closest to the breast beam, to the viewer's left.

When weft sections are aligned vertically on the page, the face of the textile is positioned to the left and the first weft is drawn at the lower end of the draft. If a weft section is drafted horizontally, the weave's face is uppermost with the first weft on the left. Choice between vertical or horizontal alignment of weft sections depends on screen or page format and how the designer prefers to record weave information.

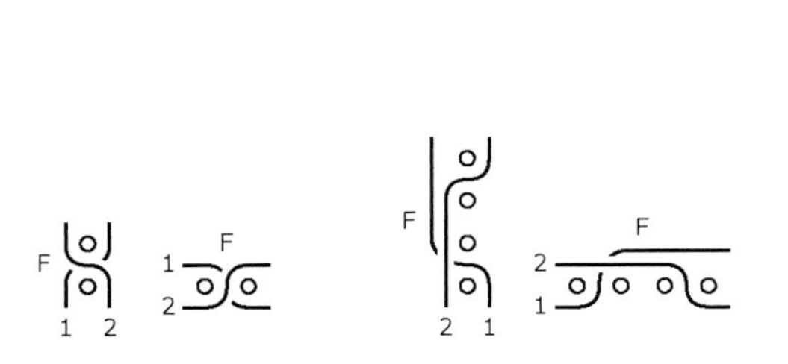

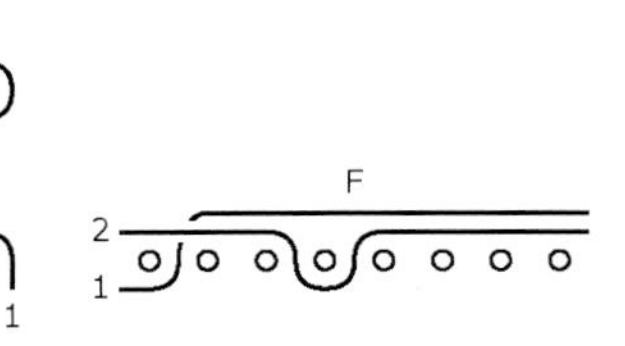

Figure 3.11. *Weft sections for each of the three weaves used in* Pomegranate *are shown aligned both vertically and horizontally.*

Observe the following drawdowns and weft sections of commonly used simple weaves. Note the number of ends and picks needed to draft each complete weave repeat.

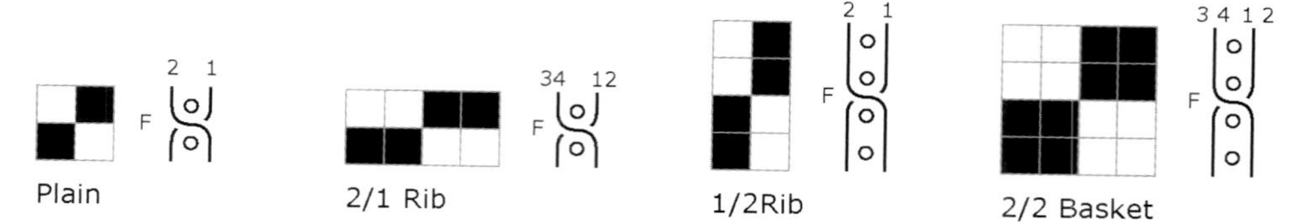

Figure 3.12. *Drawdowns and weft sections of plain weave and derivative structures.*

If two or more adjacent ends follow the same interlacement order, only one end may be drafted and the numbers of the ends indicated. See 2/1 rib and 2/2 basket in Figure 3.12.

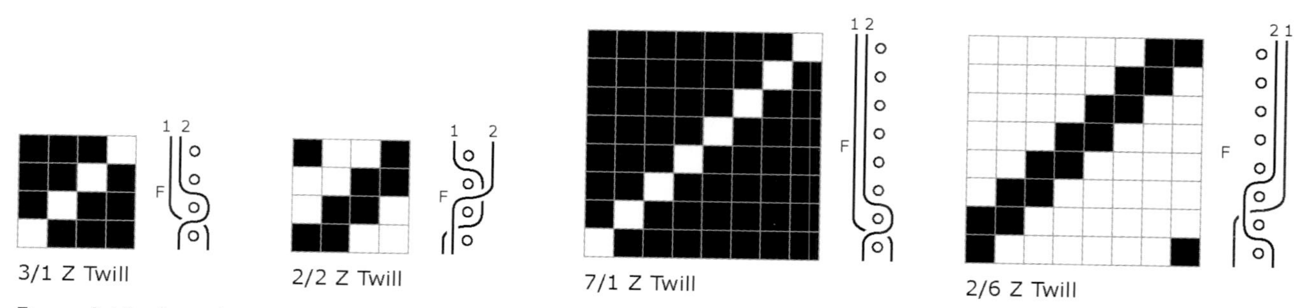

Figure 3.13. *Drawdowns and weft sections of several regular twills.*

The binding order of regular twills shifts up (or down) one point of interlacement on each successive warp end until the weave repeat is complete. When drafting the weft section of a regular twill, all picks of the repeat and only two adjacent ends are sufficient to record the complete interlacement order and direction, Z or S, of the twill's diagonal.

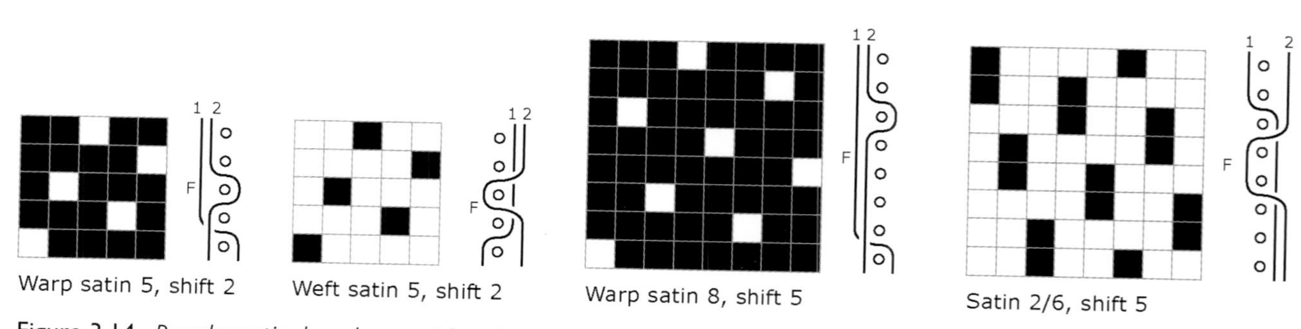

Figure 3.14. *Regular satin drawdowns with weft sections.*

The binding order of regular satins follows a fixed shift or step upward on each successive end throughout the weave's repeat. As with regular twills, a weft section showing the first two ends and all the picks of the repeat are sufficient to describe the interlacement order of regular satins.

There are times when a weft section cannot correctly record a weave, as in the case of the shaded satin 2/3 on the right of Figure 3.15. A warp section, showing the weft in profile, correctly describes the weave. Note that in this shaded satin, the transition from weft to warp satin progresses horizontally with the addition of a raised warp end next to the binding points of weft satin 5.

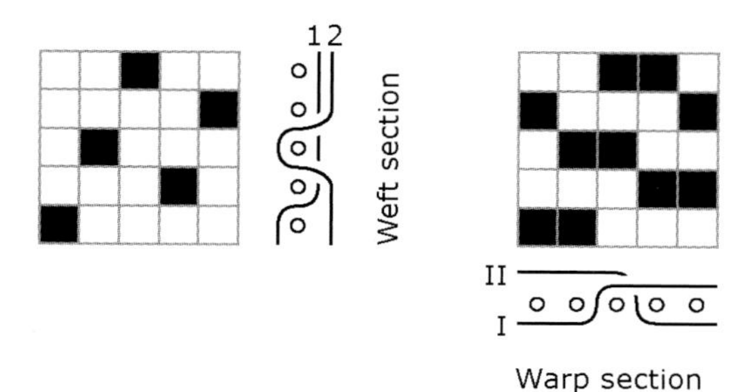

Figure 3.15. *Drawdown and weft section for weft satin* 5 on left; to the right, *drawdown and warp section for shaded satin 2/3 showing all ends and two weft picks in profile;* Right: *Photograph of cotton multi-effect damask woven with both vertically and horizontally shaded 5-end satins.*

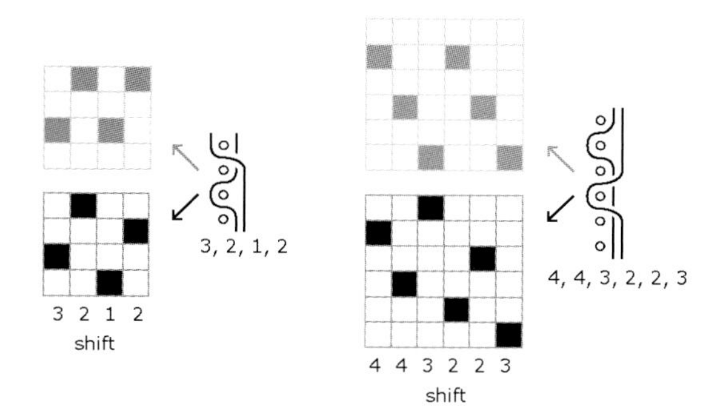

Figure 3.16. *Drawdown, weft sections, and vertical shifts for two irregular weave structures: 4-end broken twill and weft satin 6.*

Weft and warp sections are generally used for drafting regular weave structures. When used to record irregular weaves, sectional drafts may give misleading and incomplete information, in which case an alternative method must be found. In Figure 3.16, the weft sections of a 4-end broken twill and weft satin 6 do not correctly describe either weave. A note of the vertical shift of each end's binding point is added below the weft section to give complete weave information.

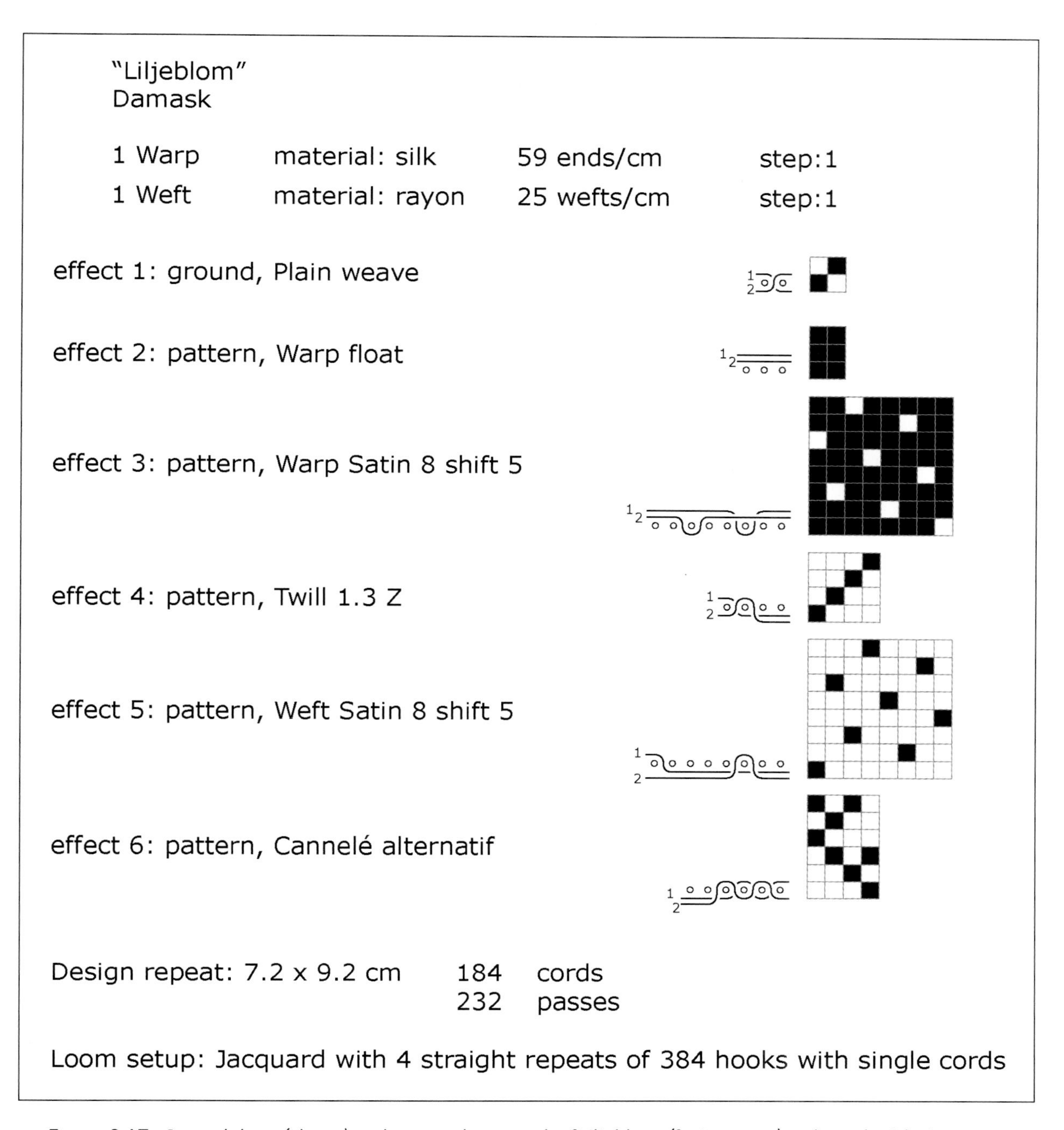
"Liljeblom"
Damask

1 Warp	material: silk	59 ends/cm	step:1
1 Weft	material: rayon	25 wefts/cm	step:1

effect 1: ground, Plain weave

effect 2: pattern, Warp float

effect 3: pattern, Warp Satin 8 shift 5

effect 4: pattern, Twill 1.3 Z

effect 5: pattern, Weft Satin 8 shift 5

effect 6: pattern, Cannelé alternatif

Design repeat: 7.2 x 9.2 cm 184 cords
232 passes

Loom setup: Jacquard with 4 straight repeats of 384 hooks with single cords

Figure 3.17. *Record sheet* (above) *and macro photograph of* Liljeblom (facing page)*, a damask with six weave effects, by Martin Ciszuk, LFS.*

Observe the record sheet for a damask with six weave effects. Each structure is listed by function (ground, pattern, binding, etc.) and weave name, then drafted as a horizontal weft section and as a drawdown on squared paper. Note that the weft section of weave effect 6, an irregular rib labeled *cannelé alternatif*, cannot be recorded solely by sectional draft; a drawdown in this case is required to record complete weave information.

Drafting compound weave structures

Representations of simple weaves (structures composed of a single series of warp ends and a single series of wefts) are easy to comprehend, as these correspond closely to the appearance of woven cloth. Compound weave structures are more challenging to read and draft. If multiple elements (warp or weft series) are drafted in sequence, the resulting drawdown looks nothing like the actual cloth. Likewise, when a compound structure is broken down into its component weave structures, these bear little resemblance to the real weave.

Illustrations, simulations, and photographs that show the appearance of the finished textile are unable to show all series of warps and wefts at the same time, as these interlace above and below each other during weaving and finishing, and ultimately may come to rest on several planes, with only the uppermost warp ends and picks fully visible.

Figure 3.18. Fish *by Veronica Tibbits, LFS. Damask with two self-patterning wefts.*
Two simple weaves, weft satin 8 shift 3 and plain weave, form the weft-faced compound structure employed for the fish motifs.
From left to right, upper row: *these are drafted a) as simple weaves in black and white, b) a compound weave in black and white, c) a compound weave using distinct colors for each element—black for the warp, grey for weft 1, pink for weft 2.*
Lower row, from left to right: *the simulation on the left shows the interlacement of all ends and picks in sequence; note that there is little resemblance between this simulation and the physical weave. The next two simulations show the appearance of the weave when compacted; however, the interlacement order of weft 2 is no longer visible.*
Facing page: *Details and 1:1 photo of* Fish.

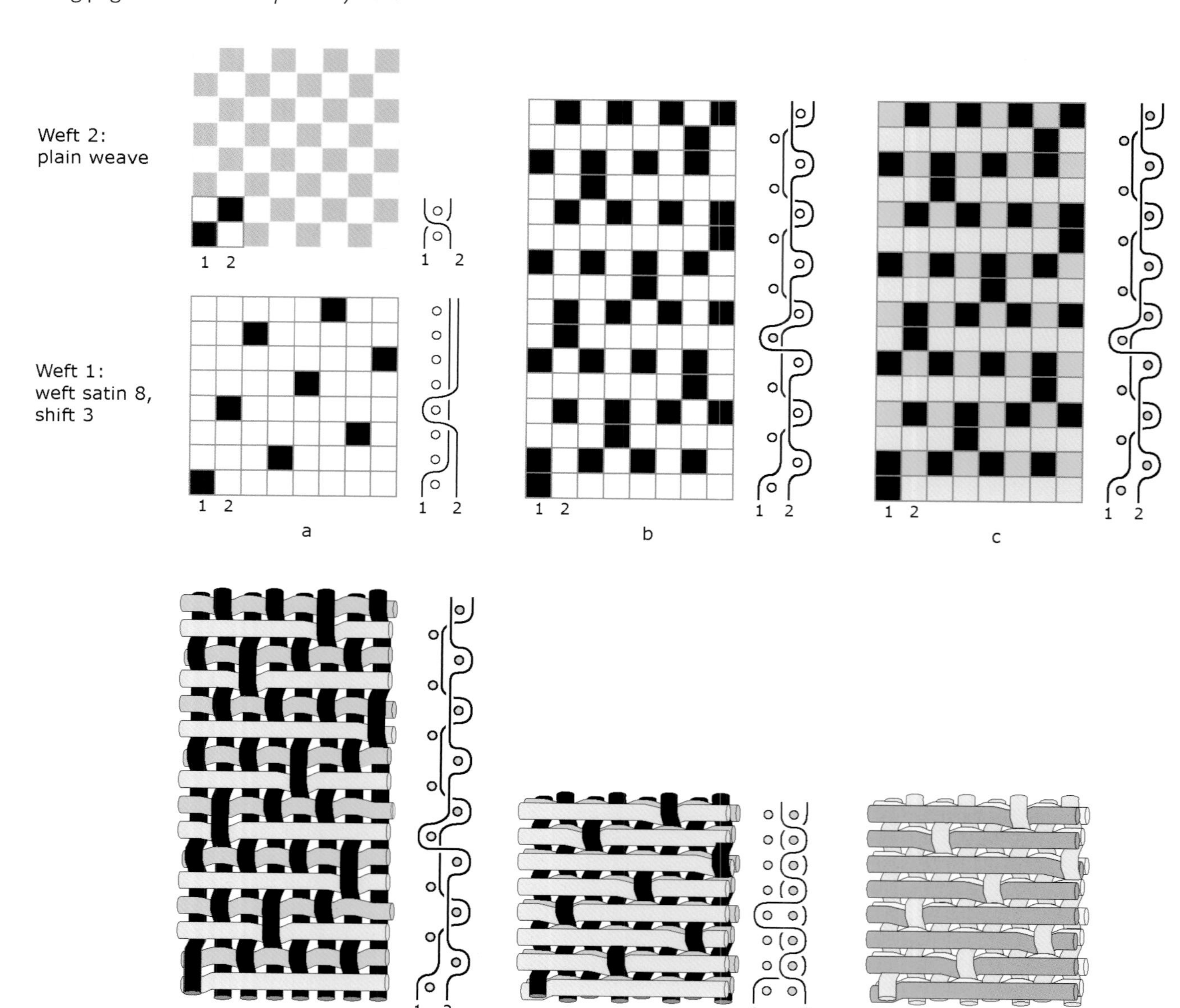

Drawdowns on Squared Paper: Whole or Decomposed

Throughout this book, two methods are used to draft drawdowns for compound weaves. The first combines all warp and weft series in a single drawdown. All warp ends are drafted from left to right in the sequence these occupy in the reed and the real textile, irrespective of function and series. All picks are drafted in the sequence these are inserted into the shed, starting from the lower end of the draft, irrespective of function and series. Each warp and weft series is drafted in a distinct color, referred to as a *technical color*. Following the convention by which a dark sign corresponds to warp up and white to warp down, dark or saturated colors are used to mark warp over weft, while white or lighter colors are employed to indicate warp under weft. The eight warp colors in Figure 3.19 are used to draft warp series from 1 to 8; the colors of the picks in the same figure correspond to weft series 1 to 8.

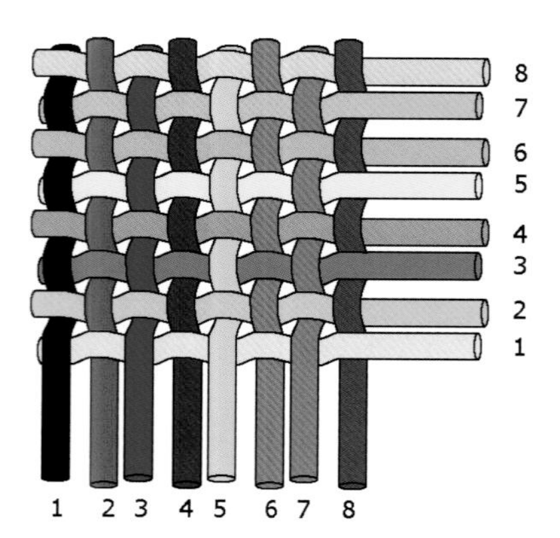

Figure 3.19. *Eight saturated colors are used to draft warp series 1 to 8; eight lighter colors correspond to weft series 1 to 8.*

Figure 3.20. Cypress Trees *by Sheetal Khanna-Ravich, LFS.*

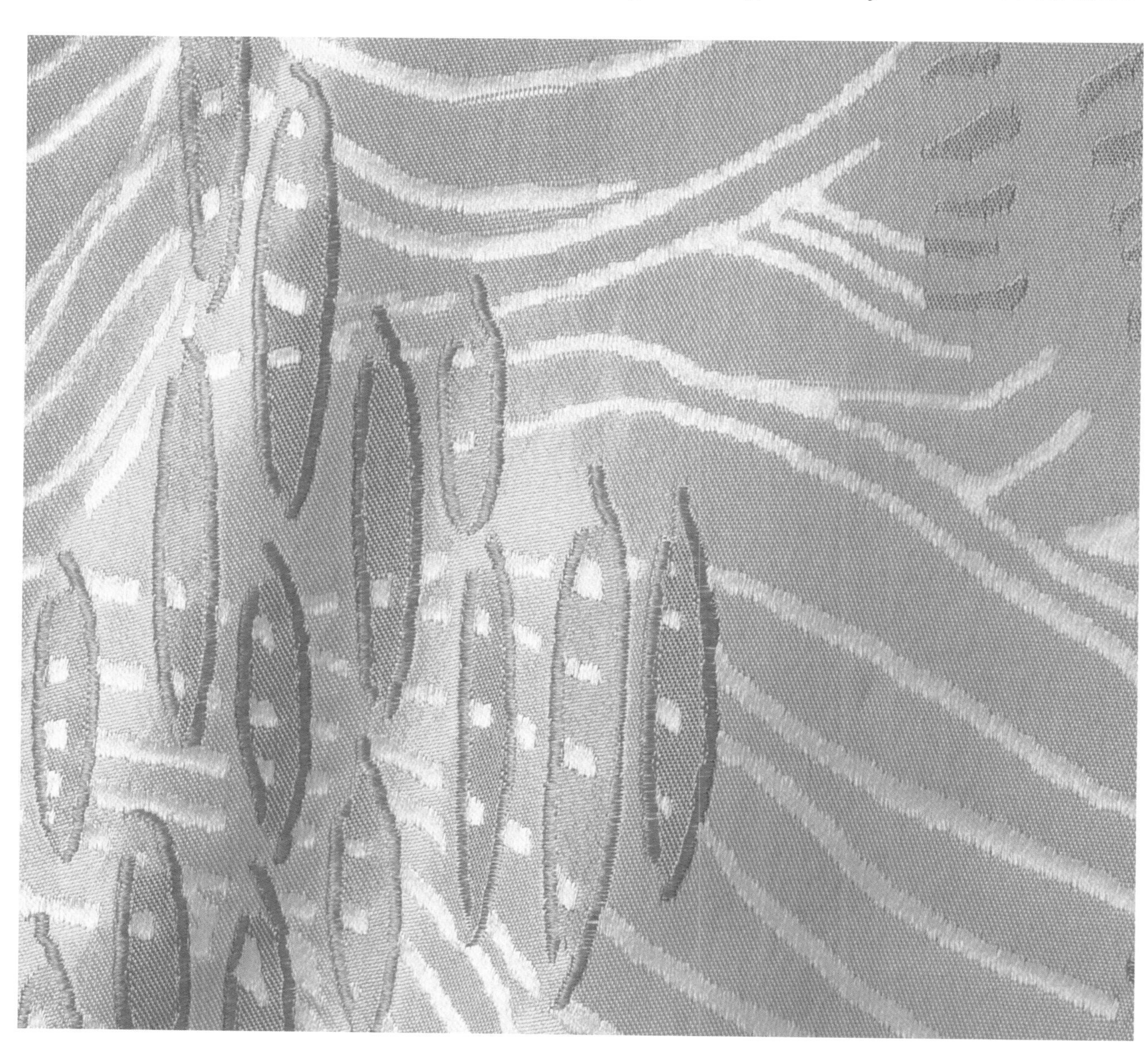

In the second method, compound weaves are decomposed into separate drawdowns for each of the simple weaves that constitute the weave structure. These may be drafted using technical colors or the black and white of conventional drawdowns.

Which drafting system is better: compound or simple drawdown? Each is useful at different stages of weave construction and textile production. When designing a new textile that employs compound weaves, it is easier to think of the interlacement of each warp and weft series one at a time, as simple weaves (see Figure 3.21). Once each simple weave is defined, textile software or traditional drafting tools are used to draft all the weaves on one plane as a compound structure. The resulting compound drawdown serves to verify weave compatibility, that is, that the binding points of each weave structure align in such a way as not to affect the interlacement of the other constituent weaves. Badly aligned or incompatible weaves not only disrupt the appearance of the uppermost weave; such errors often add unwanted take-up to one or more warp ends.

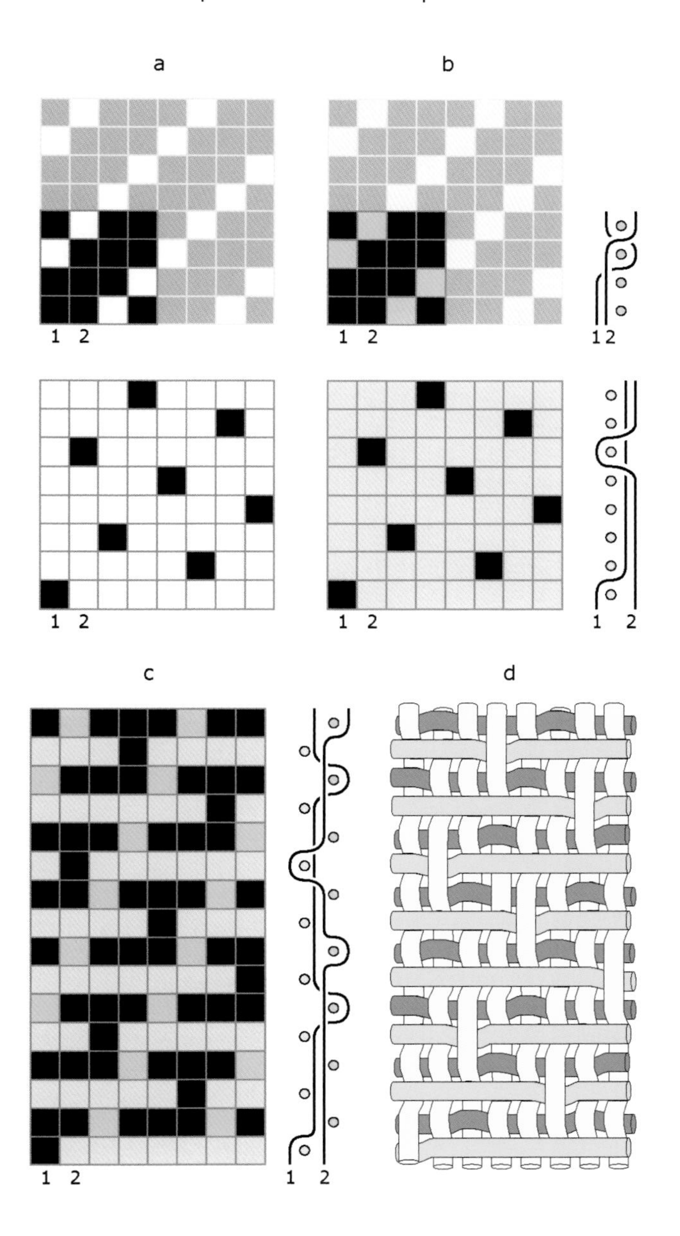

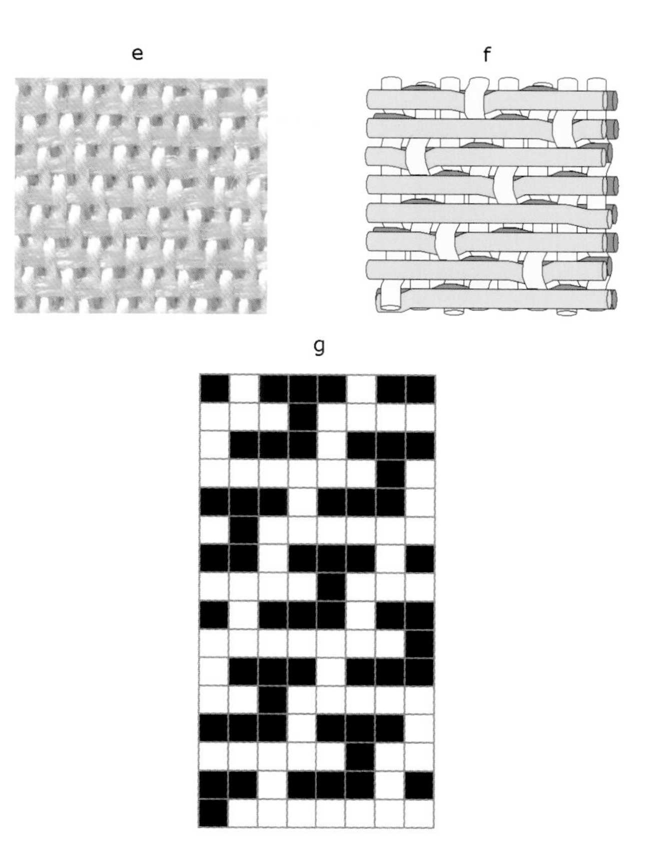

Figure 3.21. Left and above: *Drafting sequence of the ground effect in* Cypress Trees.

As corrections or adjustments are required, the compound structure is broken down into simple weaves, modified, and then reconstructed as a compound weave. At the end of the process, all ends and picks of the structure's repeat are drafted in sequential order; the position of each end relative to every pick is plotted in black for end up or white for end down. This information is transmitted to the loom's patterning device via digital interface, punched cards, levers, treadles, or pulleys and cords.

Figure 3.21 illustrates the drafting sequence of ground effect in *Cypress Trees*. In all representations, the weave is the same, that is, the first weft binds with the warp as weft satin 8 shift 5; the second weft weaves as 3/1 Z twill: a) the two weaves are drafted as simple weaves in black and white, and then b) with technical colors; c) the two simple weaves are drafted as a compound drawdown and weft section using technical colors, and then d) with the warp and weft colors; e) the weave is shown in a macro photograph and f) simulated as a compacted structure; g) the weave is drafted as a compound structure in black and white, the form used to transmit weave information to the loom.

Sectional

There are many weave patterning techniques that employ multiple warp and/or weft series. Sectional drafting is useful for visualizing the relative positions of the multiple elements used in compound weave structures.

The sections in this book are drafted with the *technical* warp and weft colors shown in Figure 3.19, or with colors that correspond to the yarns or materials of the real textile. Weft sections are used in preference to warp sections and may be drafted vertically or horizontally. As when drafting weft sections of simple weaves, two ends of each warp series are drawn in profile, as are all the picks of each weft series. If a sectional draft does not give complete or correct information relative to a weave, the section is supplemented with a drawdown or note that completes the weave record.

Often when compound structures are woven, the various elements rest on different planes, as in the figured doublecloth shown on these pages. Note the position of each series of warps and wefts in the sectional drafts in Figure 3.22.

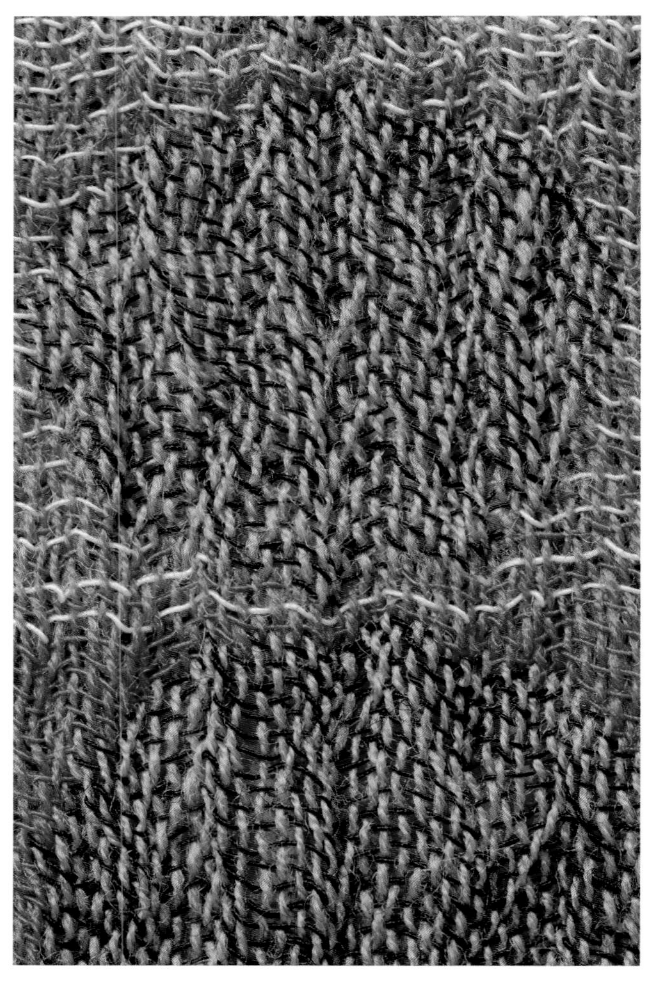

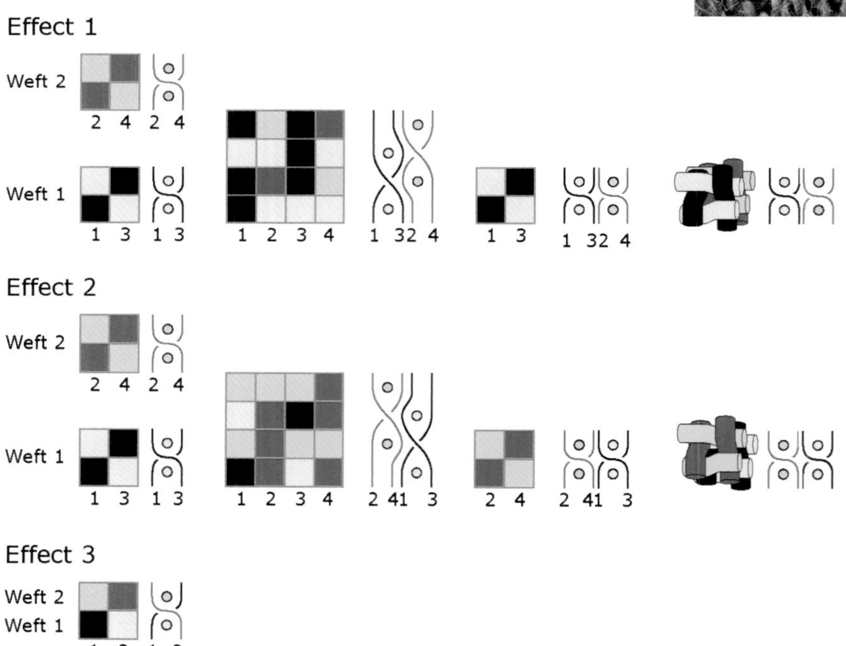

Figure 3.22. *Drawdowns and vertically aligned weft sections for the three weave effects of the adjacent doublecloth. Note that effects 1 and 2 are compound weaves, whereas effect 3 is a simple plain weave.* From left to right: *for weave effects 1 and 2: simple weft sections and relative drawdowns are drafted for the weaves that correspond to the upper and lower layers of the doublecloth, followed by compound drawdowns and sections. Next, the weave is drafted and simulated in compacted form. Note how all warps and wefts can be visualized using weft sections, whereas in the compacted drawdown and simulation the uppermost layer obscures the ends and picks of the lower layer.*
Facing page: *Photographs of the same textile, CJH.*

Technique: Lampas with Plain Weave Ground and 4 Pattern Wefts

Warps
proportion: 4 ground / 1 binding; per cm 70; material: cotton 60/2
Wefts
proportion: 1/1/1/1/1; per cm 8 (per weft); material: spun rayon

Ground:

Effect I

Pattern:

Effect II

Effect III

Effect IV

Effect V

In this record sheet for a lampas with two warp systems, one ground weft, and four pattern wefts, each weave effect is drafted as a horizontally aligned weft section. Each warp and weft is drawn with a distinct color. A reading note (see table at the end of Figure 3.23) is compiled for each of the five weave effects and describes the interlacement of every warp and weft series in each effect.

Figure 3.23. Facing page and below: *Photographs and record sheet with sectional drafts and reading note for Francesco Sala's* 5 Weft Lampas, *LFS.*

	I: ground		II: pattern		III: pattern		IV: pattern		V: pattern	
	ground warp	binding warp	ground warp	binding warp	ground warp	binding warp	ground warp	binding warp	ground warp	binding warp
○	plain	plain	plain	plain	plain	plain	plain	plain	plain	plain
○	all up	plain	all down	plain	all up	plain	all up	plain	all up	plain
○	all up	plain	all up	plain	all down	plain	all up	plain	all up	plain
○	all up	plain	all up	plain	all up	plain	all down	plain	all up	plain
○	all up	plain	all up	plain	all up	plain	all up	plain	all down	plain

Other methods for recording simple and compound weaves

Over time, many drafting systems have developed to record weaves. Variations among systems derive from the specific needs of the textile technique recorded or the equipment employed for weaving, arbitrary choices of the individual drafter or author, and drafting materials.

If weave drafts found in notebooks and books are difficult to comprehend, try converting these into polychrome drawdowns, using darker colors for warp series and lighter colors to represent weft series. If textile software is available, convert the signs used to plot warp and weft positions into filled and missed squares. Often the resulting structure produces a recognizable interlacement plan, as shown in Figure 3.24.

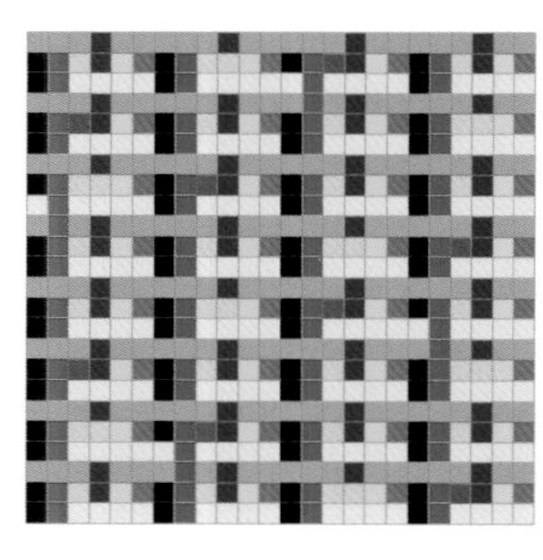

Figure 3.25. Above: *A compound weave with six warp and three weft series is converted from a black and drawdown to a polychrome draft;* Facing page, top: *photograph of industrially woven tapestry,* Sandscroll, *by Robin Muller;* Facing page, bottom: *printout of* Sandscroll*'s weaves. Courtesy of RM.*

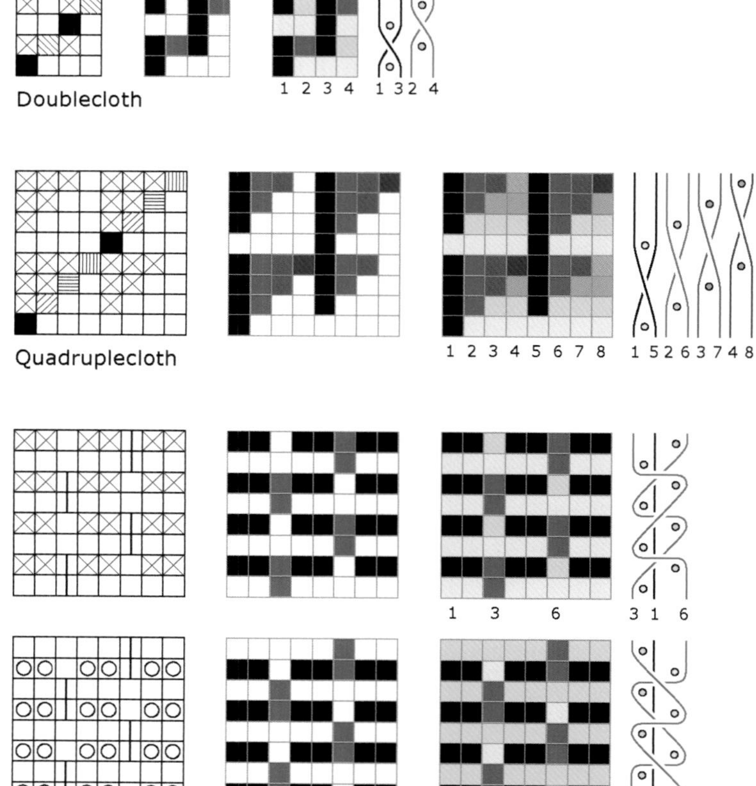

Figure 3.24. *Weave drafts that employ a variety of symbols and graphic devices to record warp and weft positions.* Left: *original draft,* right: *converted drafts.*

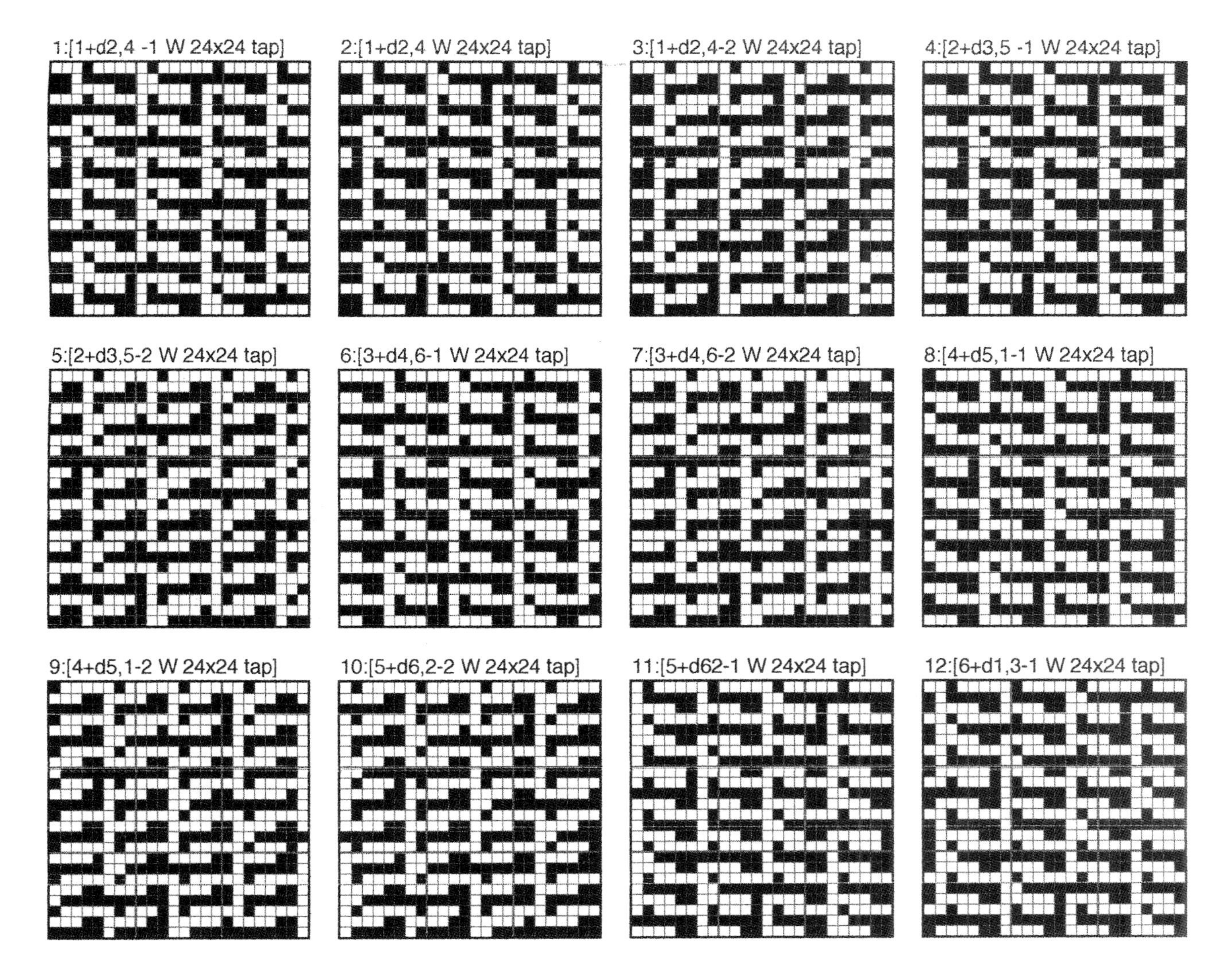
1:[1+d2,4 -1 W 24x24 tap]
2:[1+d2,4 W 24x24 tap]
3:[1+d2,4-2 W 24x24 tap]
4:[2+d3,5 -1 W 24x24 tap]
5:[2+d3,5-2 W 24x24 tap]
6:[3+d4,6-1 W 24x24 tap]
7:[3+d4,6-2 W 24x24 tap]
8:[4+d5,1-1 W 24x24 tap]
9:[4+d5,1-2 W 24x24 tap]
10:[5+d6,2-2 W 24x24 tap]
11:[5+d62-1 W 24x24 tap]
12:[6+d1,3-1 W 24x24 tap]

Part II

Technology, Process, Technique

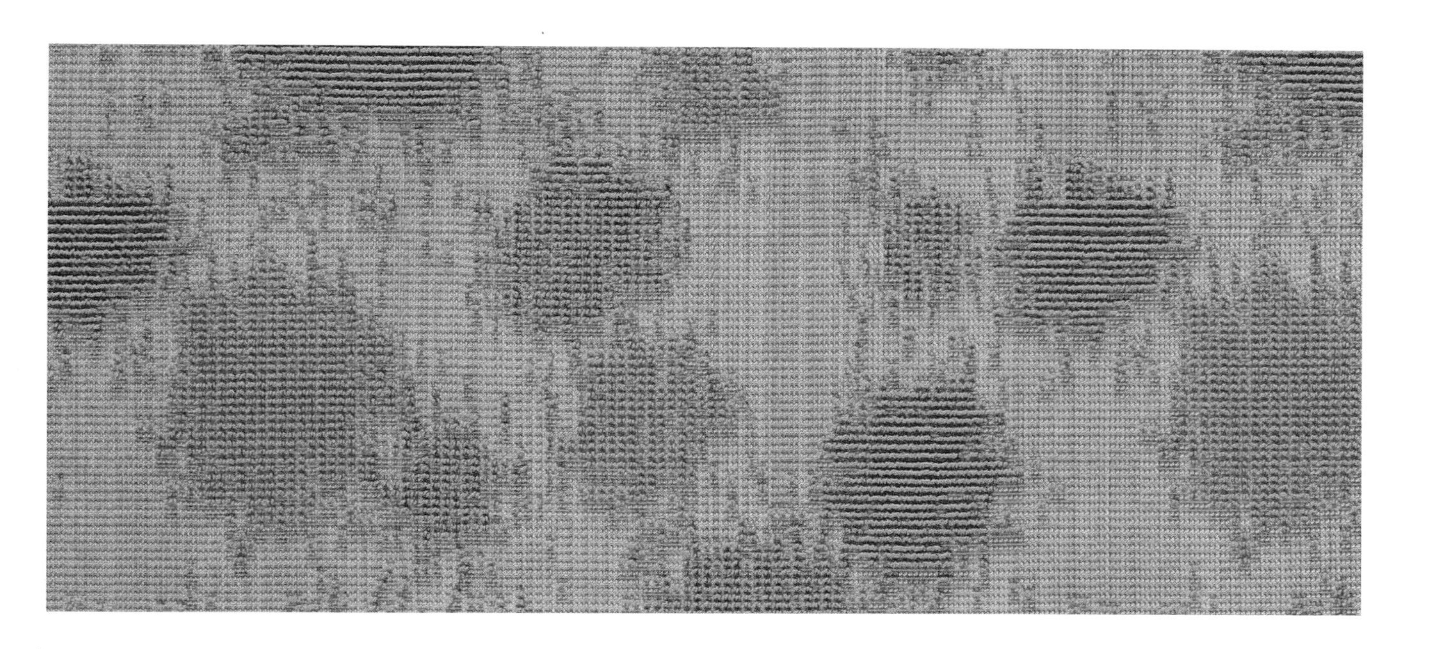

Tove's Velvet *by Tove Andéer, LFS.*

4 Weave-patterning technology in the digital era

A brief history of figuring technology

The earliest figured textiles known to man no longer exist, but come down to us through time as vestigial images of fibers, yarns, and their interlacement imprinted on metal.[1] These testimonies from before the Common Era bear the legible exchange of more than one weave within a single artifact.

In the first millennium c.e., figured textiles were produced in China, Central Asia, Byzantium, and Egypt, and arrived in Europe from the eastern Mediterranean as goods of trade and booty of war. Contact with traders and technicians spurred the establishment of silk productions in Spain[2] and Italy[3] in the second millennium. Later the technology and knowledge required to weave figured cloth spread to France and then north and east through the rest of Europe until many countries had developed their own traditions for the production of this class of wovens. The loom on which these intricate artifacts were produced was called a drawloom and is the ancestor of the Jacquard looms we use today.

All looms have much in common, starting with the tensioning of the warp. Horizontal floor looms designed to weave meterage of simple cloth are equipped with what is called a harness, or series of moving parts—heddles, shafts, and treadles or a dobby device—that lift and/or lower groups of warp ends to create an opening, called the shed.

One pick at a time, the weft is inserted into this opening and traverses the warp at right angles. A beater presses the weft toward the front of the loom until it comes to rest against a previously inserted pick. The harness again opens the shed, lifting a new set of warp ends; another pick is inserted and beaten in—woven cloth is formed. The order in which the ends and picks interlace is called the weave, the minimum unit of interlacement, the weave repeat. Variations in the lifting sequence of the shafts of the harness produce a limited number of weaves.

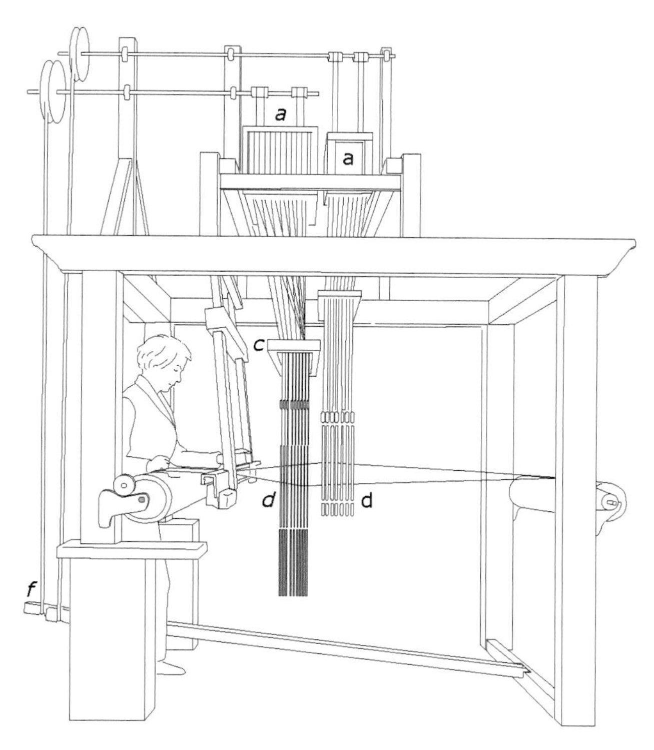

Figure 4.3. *Loom with a. Jacquard machine controls; d. figuring harness; c. comber board spreads out and maintains position of necking cords and leashes of figuring harness; a. dobby device controls d. ground harness; f. treadles activate Jacquard and dobby machines.*

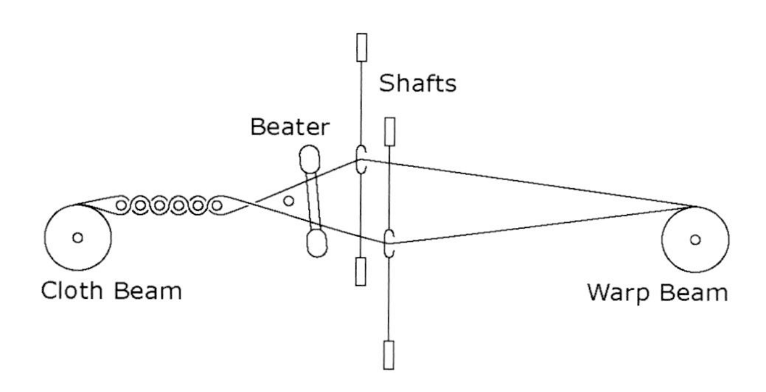

Figure 4.2. *Cloth and warp beams, shedding system, and beater of a horizontal floor loom.*

Figure 4.1. Facing page: *weft-faced compound twill, silk, Sogdiana, eighth century c.e., Cleveland Museum of Art. Photograph © The Cleveland Museum of Art.*

In a loom equipped with a figuring device, a large number of warp ends—hundreds to thousands—are lifted individually in any combination as the weft passes through the shed, vastly increasing the variety of weaves that may be woven. In drawloom weaving of the past, and with early Jacquard productions, ground and figuring harness wove in tandem. The figuring harness produced the intricate patterning, while the ground textile was woven with a simpler shaft assembly.

The principal difference between draw and Jacquard looms is found in how warp ends are selected and lifted. The first and older system requires a weaver to operate the ground harness and throw the weft while a second worker, called the draw or pulley boy, transmits the pattern to the warp by operating a complex set of cords and pulleys, called the *simple*, that raises selected ends.

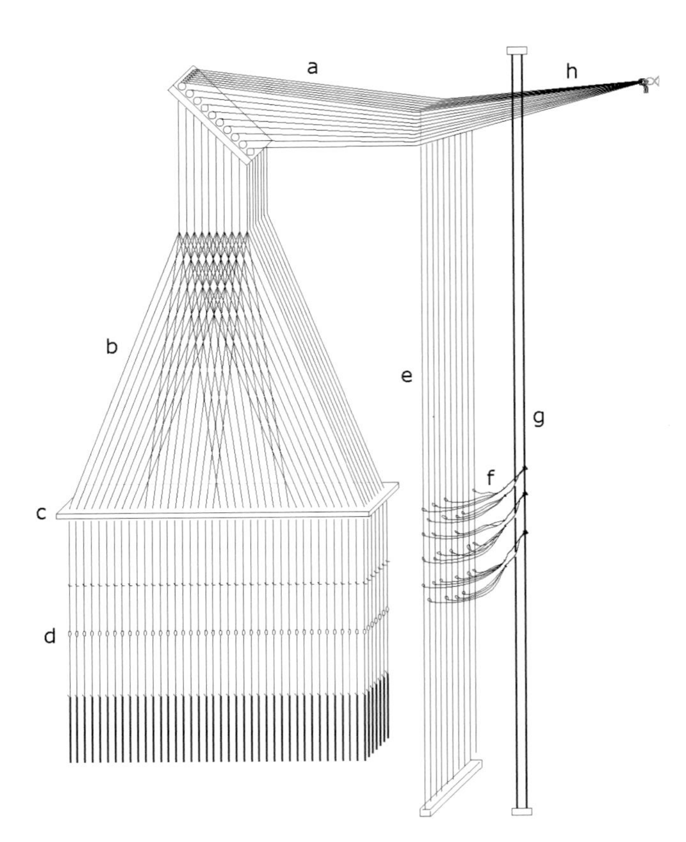

Figure 4.4. *Drawloom figuring harness: a. pulley cords, b. necking cords, c. comber board, d. leashes, e. simple cords, f. lashes, g. guides, h. tail cords.*

With the advent of M. Jacquard's invention at the beginning of the nineteenth century, control of the patterning passed to the weaver, making the drawboy superfluous. The weaver activated the Jacquard machine that read prepunched *cards* (see Figure 4.5.e) containing the pattern, and then selected and raised warp ends, while the weaver continued to throw the weft and operate the ground harness.

The monture, the ensemble of parts that compose the figuring harness of both draw and Jacquard weaving, is shown in Figures 4.4 and 4.5. The *comber board*, an essential part of this ensemble, spreads the necking cords out to the width of the pattern unit; the necking cords are entered in a precise order in the holes of the comber to correctly replicate the direction of the repeat, as shown in Figure 4.6.

Transmitting the pattern to the warp with either figuring device was a lengthy process; in the case of drawloom technology, a *reader* read, row by row, a *pointpaper* (a scaled design on squared paper) and tied the elaborate sequence of *lashes* (see Figure 4.4.f) that would be pulled to select the ends to be raised or lowered as each pick was thrown. To transmit patterning information to Jacquard's machine, the pointpaper design was read by a technician who punched holes in a series of cardboard cards, one card per weft shot, one hole for each warp end to be lifted. The cards were then tied into a chain and read by the Jacquard machine as each pick was inserted into the shed.

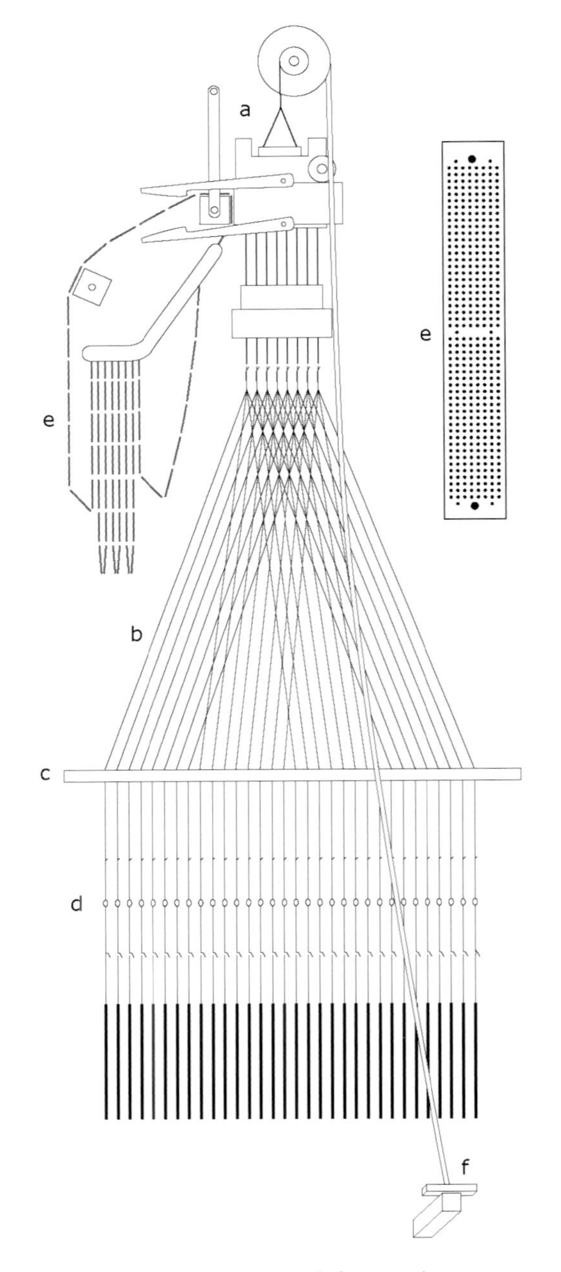

Figure 4.5. *Jacquard machine with figuring harness: a. machine, b. necking cords, c. comber board, d. leashes, e. cards, f. treadle.*

Over time the force required to activate the Jacquard device was supplied by machinery, and weft insertion became mechanical. Gradually the transmission of patterning information to the warp became more efficient, with a capacity to control an ever-increasing number of ends across the loom's width. As the number of programmable ends grew, the need for a separate figuring harness was surmounted, and the ground harness in industrial weaving became a rarity. Ultimately the Jacquard harness wove both figure and ground.

In the last generation of industrial Jacquard looms, the patterning information is no longer transmitted via prepunched cards or paper. Electronic interfaces send design information to minute solenoids that select warp ends that are then raised by machinery able to weave more than five hundred picks per minute, while the weaver monitors multiple looms at one time, operating the digital controls that transmit production data to each loom.

Figure 4.7. *A mechanical Jacquard loom mounted with a single harness that controls ground and patterning.*

The digital loom comes down; the comber board goes out

The digital era has revolutionized the tools and processes of Jacquard weaving. Soon the 200-year-old name we use today for figuring technology may become obsolete, like that of so many inventions from the past. Increasingly both industrial and handlooms are equipped with a selecting/lifting mechanism that controls each end of the entire warp, eliminating the repeat and need for a harness and comber able to transmit multiple units of the design to the warp.

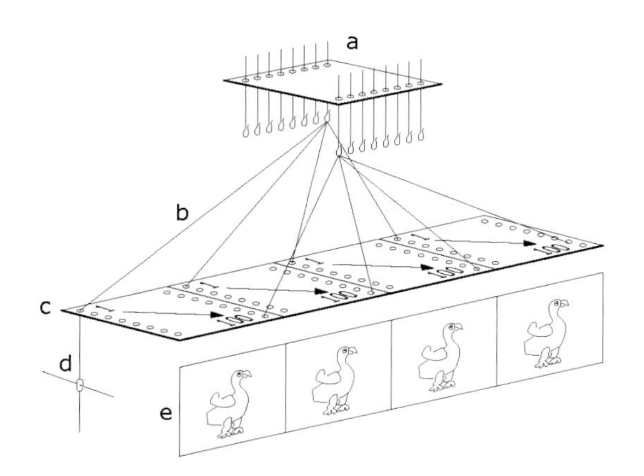

Figure 4.6. *Necking cords entered straight in four comber units: a. neck board, b. necking cords, c. comber board, d. leashes, e. four repeats of pattern unit.*

The digital loom is "coming down." The era of the tall loom that accommodated a harnessed repeat and vast mechanical device is drawing to a close. With the elimination of the comber, figured weaving will gain the flexible warp density of shaft and dobby weaving. The once unaffordable and difficult to access figuring loom has become commonplace in university labs and artists' studios, while industry develops looms that will bring down the roof once raised by M. Jacquard's invention.

Not only the loom has changed: weave software allows artists and small- and large-scale productions to create designs rapidly, assign weaves, and instantly transmit the vast body of data—the position of thousands of warp ends relative to as many picks—to digital equipment that controls warp action. This abbreviation of the phases that precede weaving also facilitates experimentation. Artists and industrial designers can construct custom weave structures, modify design and technique, verify with simulation, and then sample new ideas rapidly. Not only innovative structures and techniques can be used; a rich heritage of past techniques, too labor intensive to survive the industrial revolution, is now viable again in an era of easily transmitted weave data.

Industrial or manual Jacquard weaving: fast or slow

As the entire cycle of figured textile production adopts digital technologies, the time it takes to weave cloth is little changed. Ultimately, the difference between the slow and fast weaving of manual or industrial productions remains.

Weaving speed determines what materials may be used for warp and weft and how these interlace. Fragile materials, irregular warps and wefts cannot support the high warp tension and lightning-fast weft insertion of industrial weaving. To date, precious metals in lamella form, unusual wefts such as feathers and leather, and many "intelligent" materials can only be woven at the slower speeds of handweaving. The relaxed warp tension of handwoven cloth lends depth to the interlacement of warp and weft, and a hand-thrown shuttle need not travel from selvage to selvage, and may be used to brocade. As never before, weave-patterning has become a medium accessible to artists, designers, and industrial weavers.

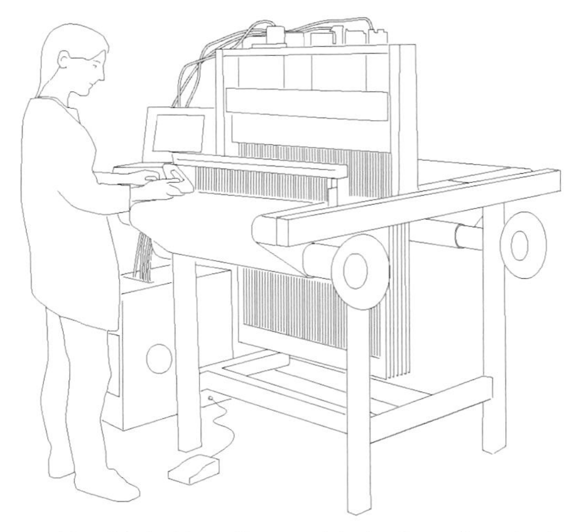

Figure 4.8. *Digital handloom with single end control, without harness or comber board.*

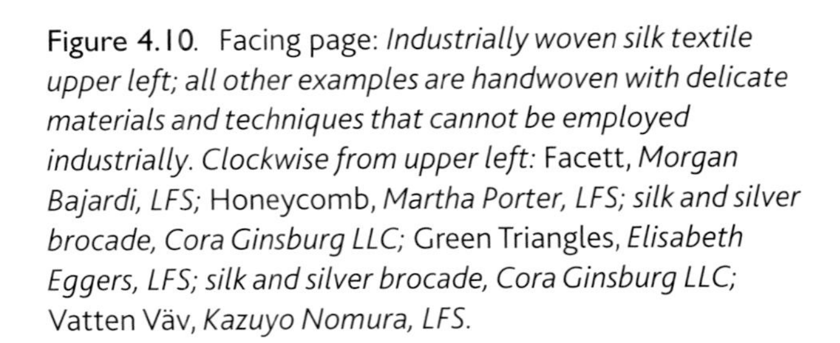

Figure 4.10. Facing page: *Industrially woven silk textile upper left; all other examples are handwoven with delicate materials and techniques that cannot be employed industrially. Clockwise from upper left:* Facett, *Morgan Bajardi, LFS;* Honeycomb, *Martha Porter, LFS; silk and silver brocade, Cora Ginsburg LLC;* Green Triangles, *Elisabeth Eggers, LFS; silk and silver brocade, Cora Ginsburg LLC;* Vatten Väv, *Kazuyo Nomura, LFS.*

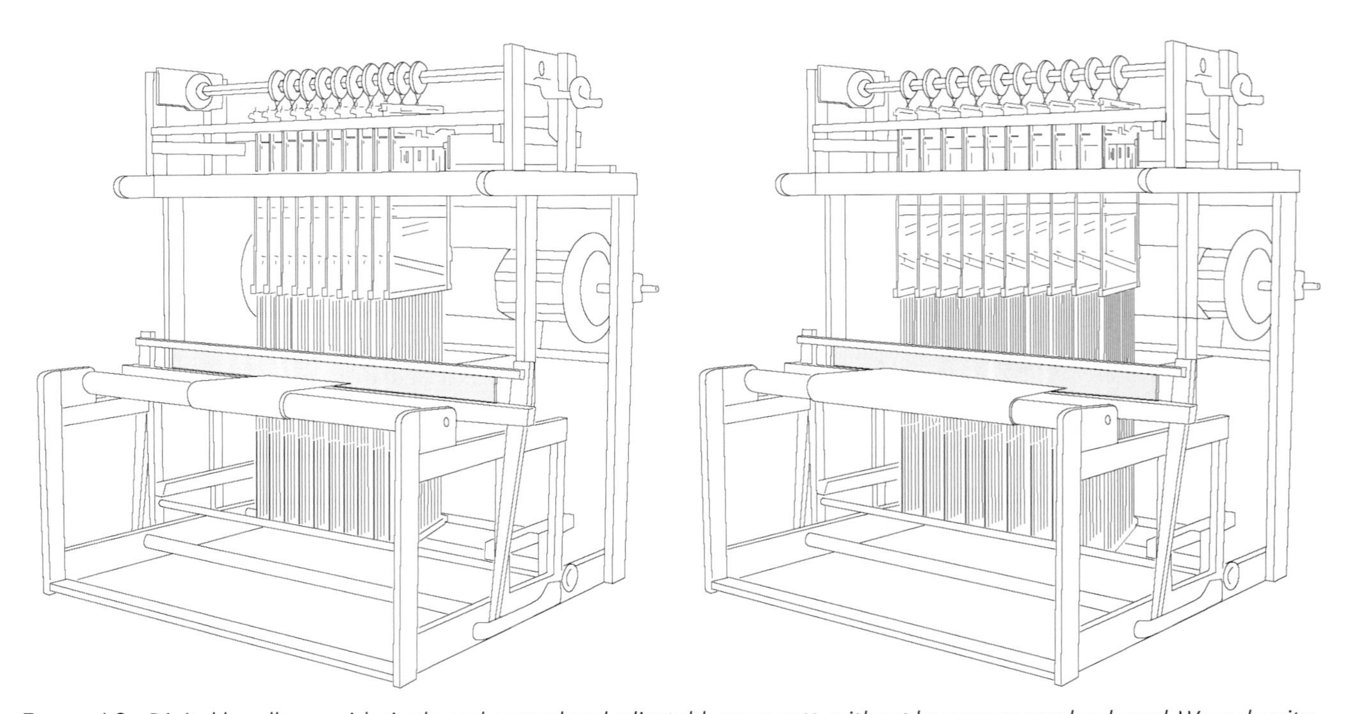

Figure 4.9. *Digital handloom with single end control and adjustable warp sett, without harness or comber board. Warp density adjusted with accordion system containing end controllers.* Left: *high sett,* right: *low sett.*

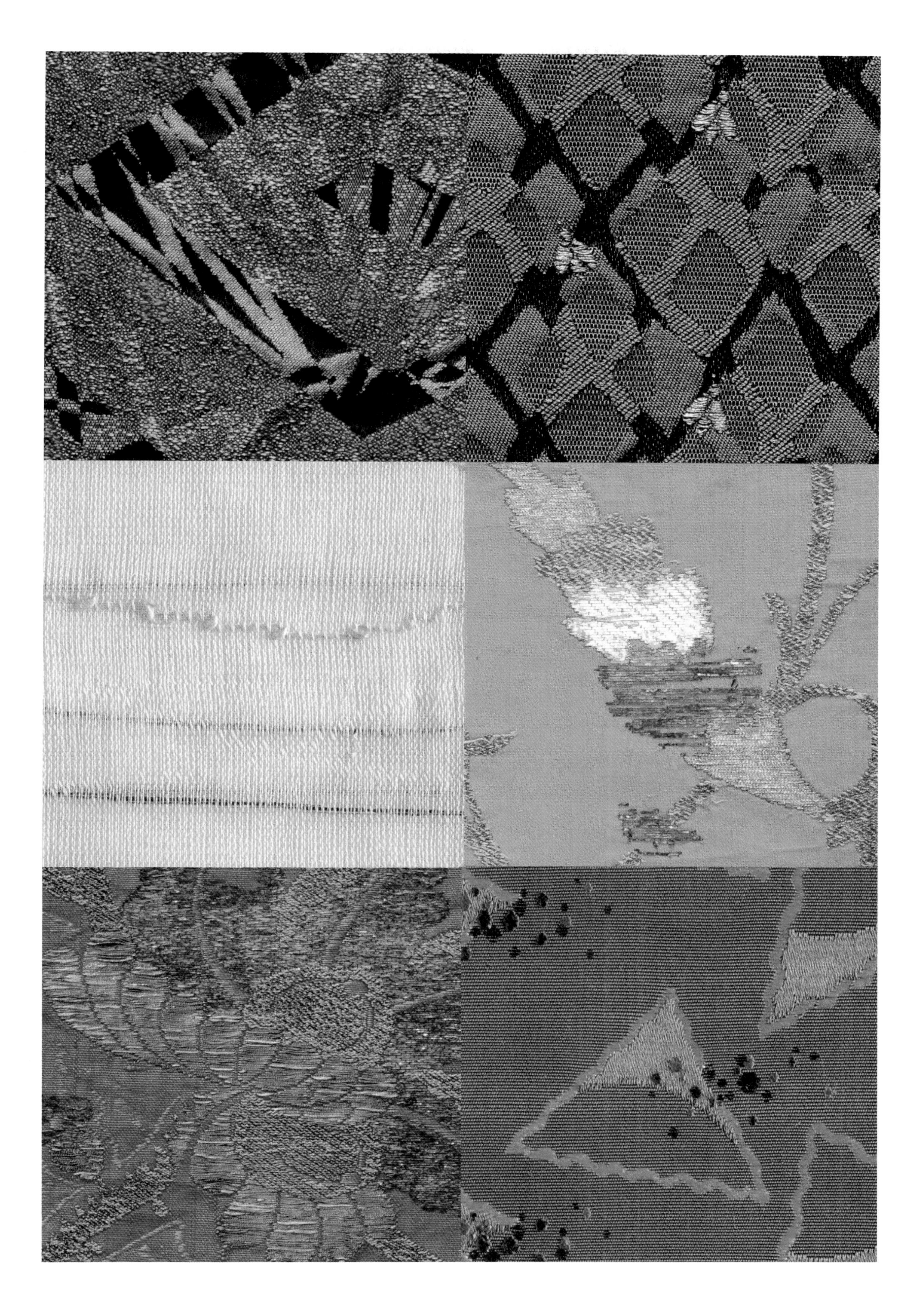

5 Design process and terminology, from sketch to card

Weave-patterned textiles have been woven for thousands of years. The technologies used to produce them evolve continually, building on age-old knowledge passed on from one generation to the next. Today, digital technology aids in the creation of the design and controls the loom, yet many of the steps from the designer's first inkling of intent to achieving the finished textile have remained the same.

The steps in current Jacquard design practice may not always occur in the same order, but typically include: definition of loom and warp/weft specifics, development of artwork, transformation of artwork into a technical design, choice of weaves, digital sampling, correction, transfer of patterning information to the loom, and production of a woven sample.

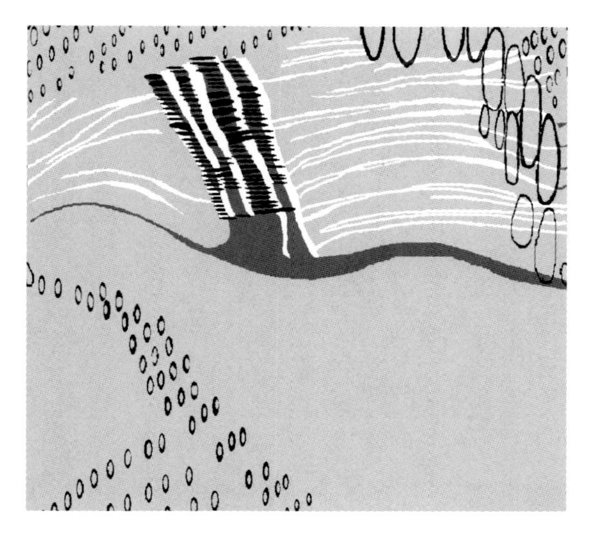

Figure 5.2. *Motif extracted, colors reduced.*

In order for weavers to communicate with figured textile producers, read new and old literature on the subject, and use professional software and tools, they must understand the process and "language" of figured textiles. Terms associated with weave-patterning processes vary widely, depending on many factors such as fiber content. Silk, hair and wool-based, cellulose, synthetic: all require specific expertise, tools, and relative terminology, which have evolved in diverse periods and geographical areas. With the advent of today's compact and affordable handlooms with figuring capacity and a choice of software, new words for figured weaving processes have proliferated. Which terms are used in this text? Terms have been selected from a variety of sources, all of which are found in widely used texts and/or in industrial practice.

Figure 5.1. Facing page: *preliminary sketches for* Cypress Trees, *drawn and painted with traditional means.*

Process

Defining the Project

The first step in the design process is to establish the specifics of two factors that dictate every project: loom setup and warp. The loom dictates the textile's width, the patterning device the design's width. A combination of loom and warp characteristics determines the techniques that can be woven: warp sett, warp sequence and materials, and the number of warp beams are the most important of these, while other factors may expand or limit the type of weave structures that can be woven using a given loom and warp. Weft materials, sett, and sequence are variable and are determined by the availability of materials and by commercial factors.

Artwork, Design

Based on the known parameter of image width (determined by the loom setup) the maker generates artwork or a design using any combination of traditional drawing or painting tools, photography, or graphic software.

Repeat

If the loom is set up to weave more than one repeat of the design, then the designer decides if the image must connect at the edges to the adjacent repeat, and if so, adjusts the image to meet up at the boundaries of the motif.

Figure 5.3. *The motif in repeat, adjusted to connect smoothly to identical repeats placed on any side of the original.*

Pointpaper Design

The finished artwork is transformed into a traditional or digital pointpaper, the first and key draft of a Jacquard file. The pointpaper contains one repeat of the artwork, colors that represent every distinct weave structure used in the textile, as well as many other details of the project. The designer either traces the artwork onto a traditional squared paper and fills in the colors that represent each weave structure or converts the artwork into a digital file in which variously colored pixels are the equivalent of squares of a traditional pointpaper.

Reading Note

The designer then assigns weave structures to each color of the pointpaper. If working on traditional paper, the designer compiles a table of weaves, called the reading note, that links each color to a weave effect; if the weaves are defined with the aid of digital technology, each color is associated to simple or compound weaves that are then saved as part of the Jacquard file.

Simulation

Makers using textile software define the remaining drafts of a Jacquard file, such as warp and weft colors, yarn construction and dimension, and so forth. When the file is complete, the designer projects a preview or simulated textile on the screen. This virtual textile serves to verify the appearance of the textile before sampling. Modifications can easily made at this stage with little expenditure of materials and no loom time.

Technical Evaluation

Aided by yet other software functions, the designer controls and modifies weaves and pointpaper as needed to produce a textile that will weave correctly. Typically this evaluation includes a review of the take-up factor of each weave (how much warp shortening an ensemble of weaves generates); if there are significant differences in take-up, the total area and placement of weave effects in the overall design is reviewed, as these may augment or reduce take-up differences. Other factors may be evaluated, such as stability, resistance to abrasion, and so forth.

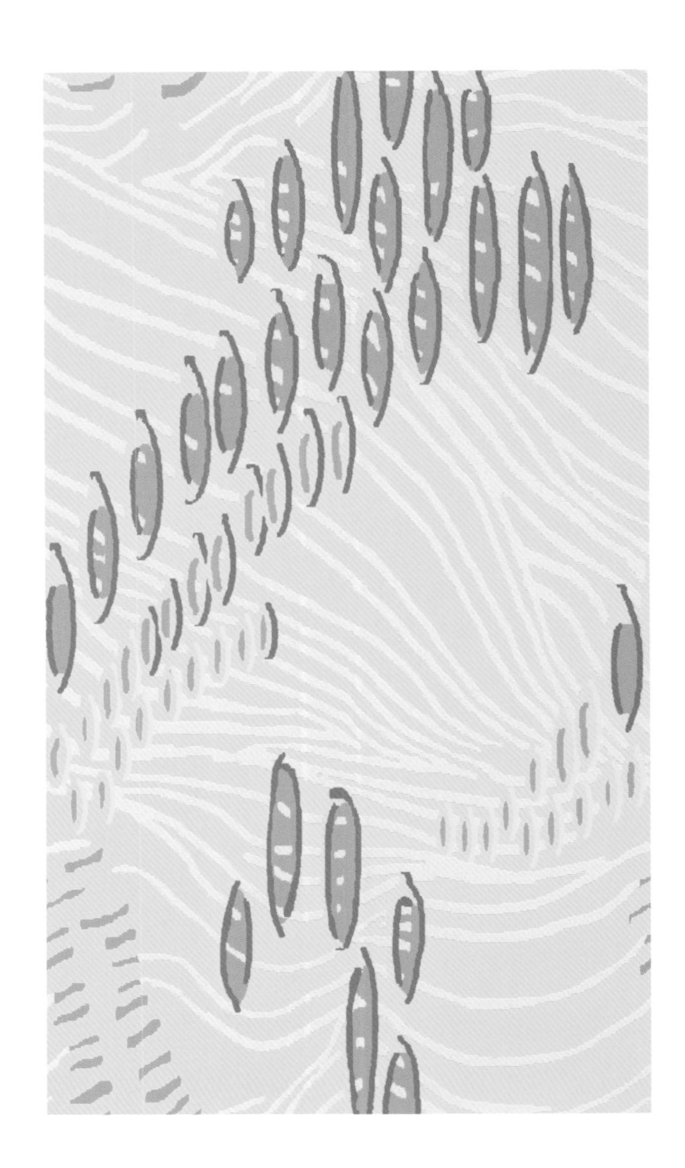

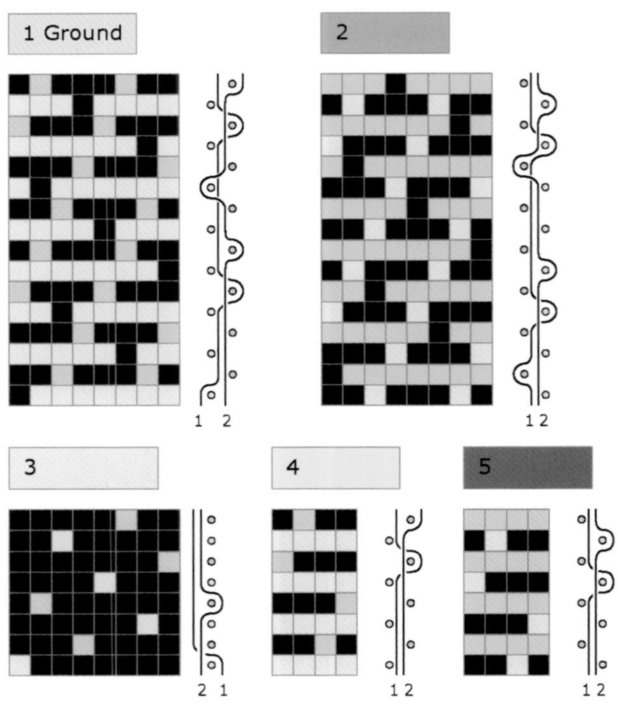

Figure 5.4. *Digital pointpaper for* Cypress Trees *drawn with five colors that represent five weave structures. Drafts of the five weaves.*

Sampling

Test samples serve to verify both the appearance and technical correctness of a project. The relative positions, up or down, of each end and pick are transmitted to the loom's figuring device by means of pulleys and cords, punched cards, or digital devices, and a portion or all of the project is woven. As needed, corrections are made to any one of the components of the Jacquard file: pointpaper, weaves; weft materials, sequence, and sett. Further samples are woven until the desired result is achieved.

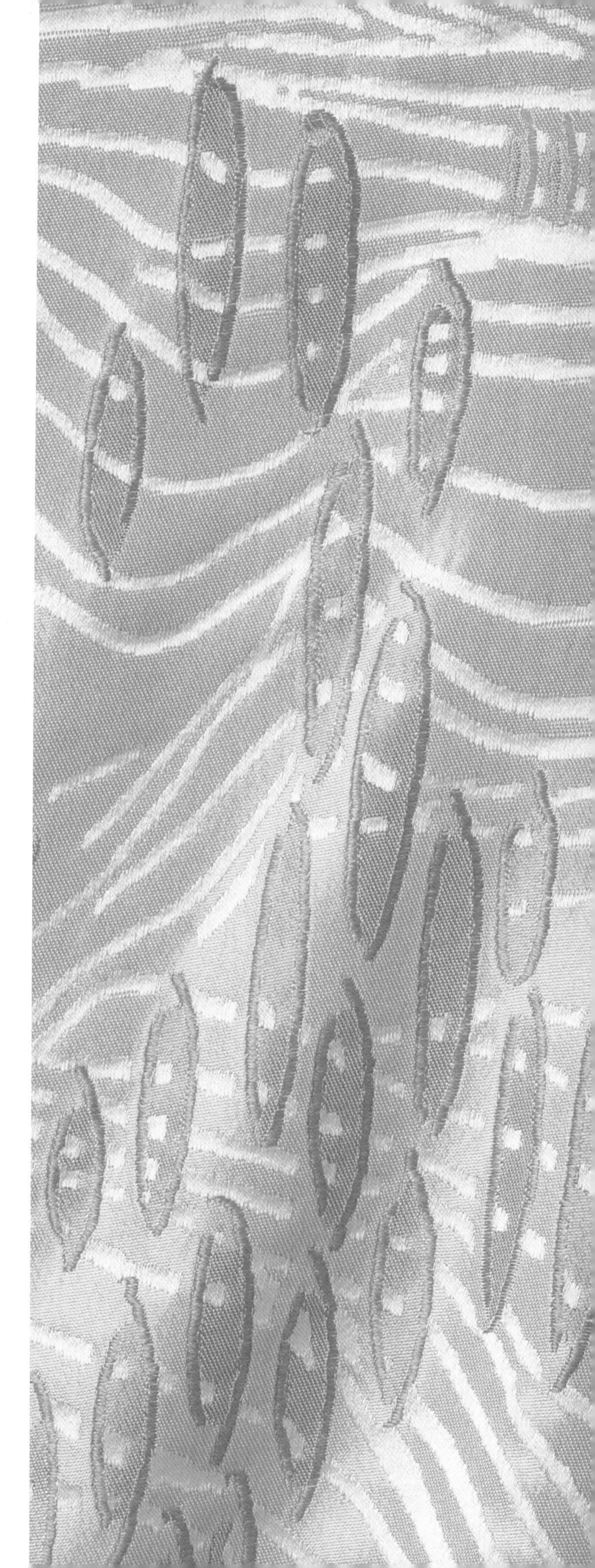

Figure 5.5. *Detail from simulation; industrially woven figured silk textile with one warp and two self-patterning wefts,* Cypress Trees *by Sheetal Khanna-Ravich, LFS.*

Terminology

The *Design* or *Artwork*: whether developed on paper or in digital format, hand drawn or computer generated, the artwork is the complete image, ready to be transformed into the pointpaper and woven.

The *Pointpaper* or *pointpaper design* is the artwork transformed into a technical draft, which may be drawn on traditional squared paper or in digital format. It is the first and key draft of the Jacquard file; it links artwork to warp and weft sequences, sett, and colors; reading note and weave effects; and finally to the loom via the card.

The pointpaper must:

1. Contain one complete repeat of the artwork.

2. Represent every visible point (or groups of points) of patterning on the textile's face with the minimum design unit: one square or pixel.

3. Represent each weave structure or *weave effect* with a distinct color.

4. Reflect the ratio of warp to weft setts via the proportions, or *aspect ratio*, of each square or pixel of the pointpaper.

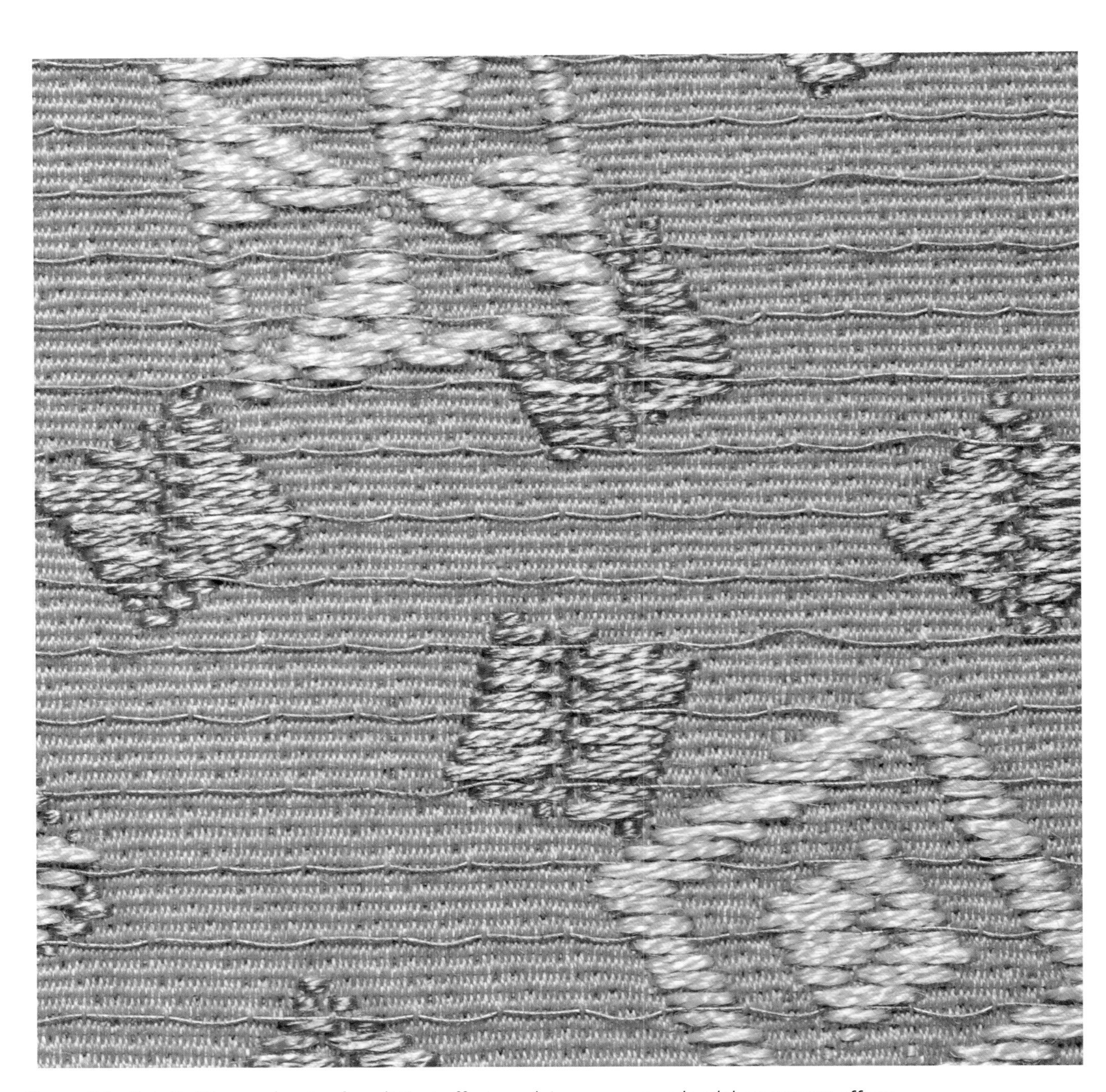

Figure 5.6. *Detail of* Boxes, *showing four distinct effects: a plain weave ground and three pattern effects.*

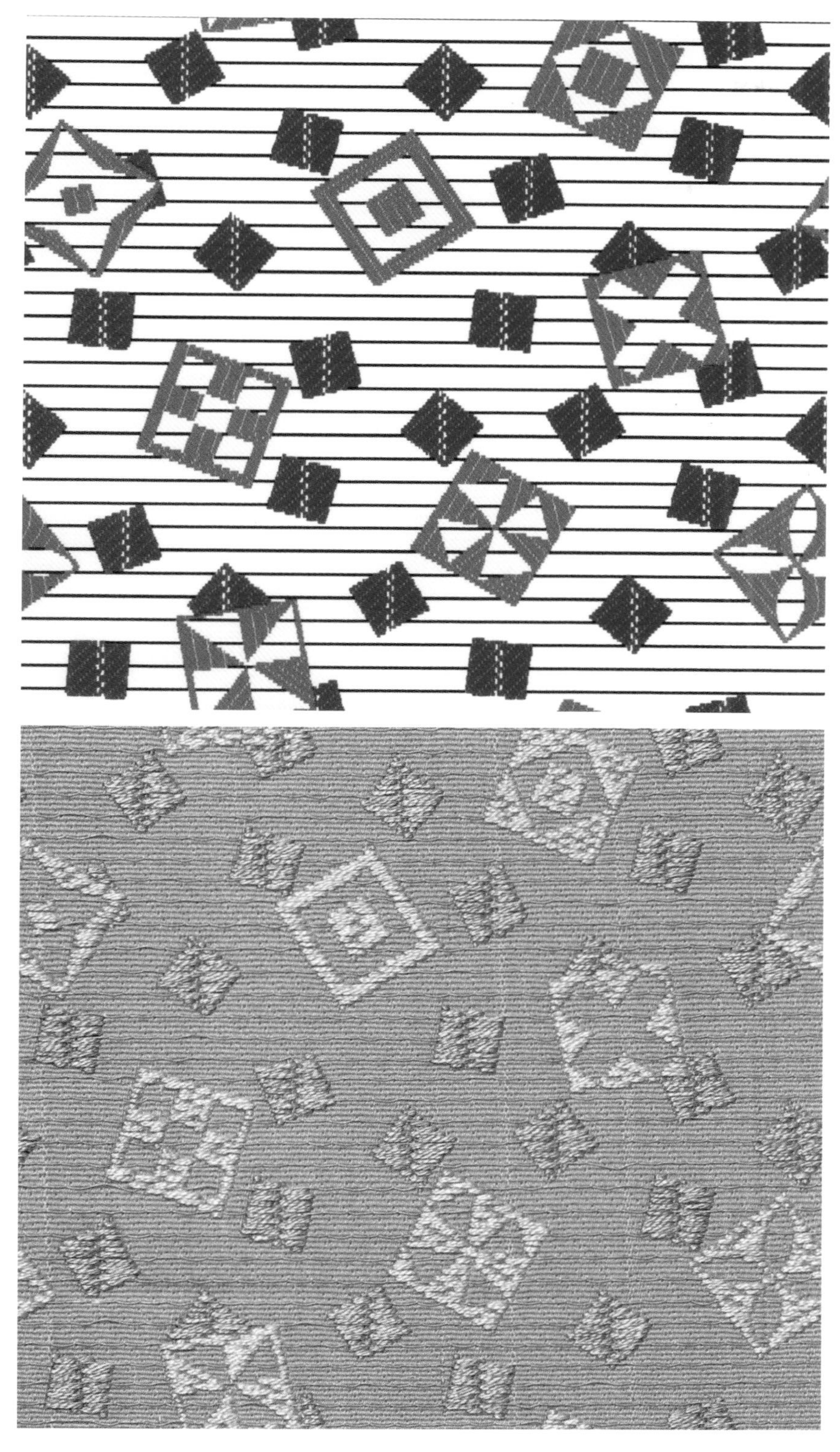

Figure 5.7. *Computer-generated pointpaper drawn at real size with four colors that correspond to the four effects visible on the surface of* Boxes. *Figured silk textile with one ground and three pattern wefts, by Joy McGruther Alaoui, LFS.*

When drawn in digital format, a pointpaper represents one repeat of the finished textile at a 1:1 scale, that is, the pointpaper design and completed textile are equal in dimension. When working on details of the image, a digital pointpaper may easily be enlarged or reduced to aid the designer as he or she evaluates the appearance of a line, motif, composition, or repeat.

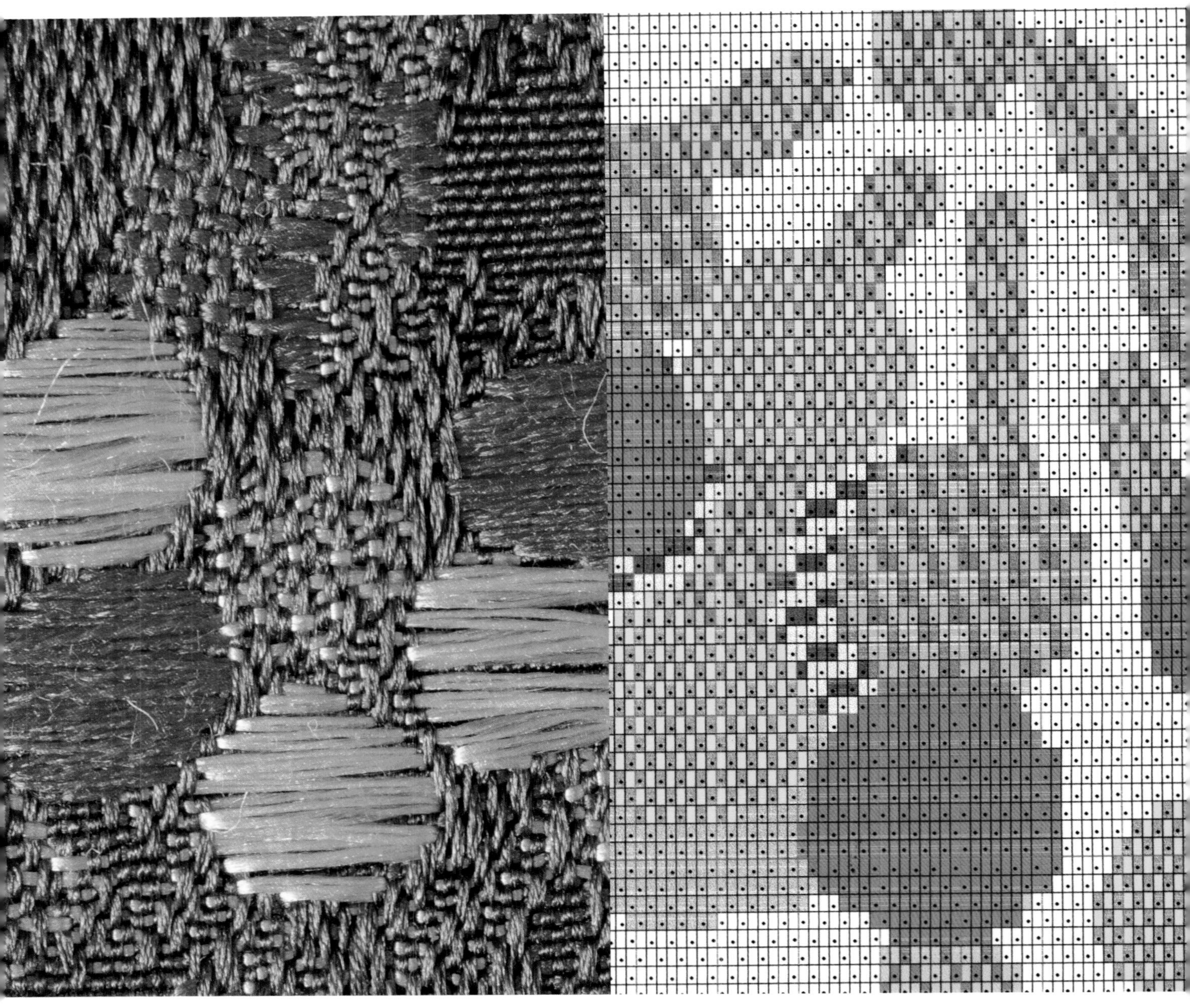

Figure 5.8. Pompoms: *detail of textile, and relative pointpaper printed on traditional squared paper. The pointpaper is much larger than the actual textile, shown at real size in Figure 5.10.*

If a pointpaper is drawn or printed on traditional squared paper, the scale of the squares must be blown up to a size that permits the designer to see each square. The enlarged design may bear little resemblance to the finished textile. The colors of the pointpaper in Figure 5.8 define the various weave areas, but do not correspond to the warp and weft materials seen in the finished sample. Bright, easy to distinguish colors are often used to draw a pointpaper, especially if the textile will be woven on a card-driven loom, as clear colors are easier to read when punching cards.

Aspect ratio: Warp and weft setts determine the area (width and height) that each point of interlacement occupies on the surface of a textile. The squares of the pointpaper represent these points, or groups of points. The ratio of the width to the height of these squares is the aspect ratio. N.B: By convention the "squares" of a pointpaper design may be rectangular, that is, non-square.

Correct definition of the aspect ratio ensures a perfect correspondence between the pointpaper design and the finished textile. If the aspect ratio is incorrectly defined, the textile's motifs and repeat will be shorter or longer than the intended result, that is, the design will be deformed, as demonstrated in the second row of Figure 5.9. From left to right: based on the aspect ratio 1:1, the circular motif of 20 x 20 squares on the left maintains correct proportions, whereas the two motifs to the right with aspect ratios of 1:2 and 2:1 become deformed. In the bottom row of Figure 5.9 the two motifs' proportions are correct when the total number of weft lines is adjusted to reflect the aspect ratios of 1:2 and 2:1.

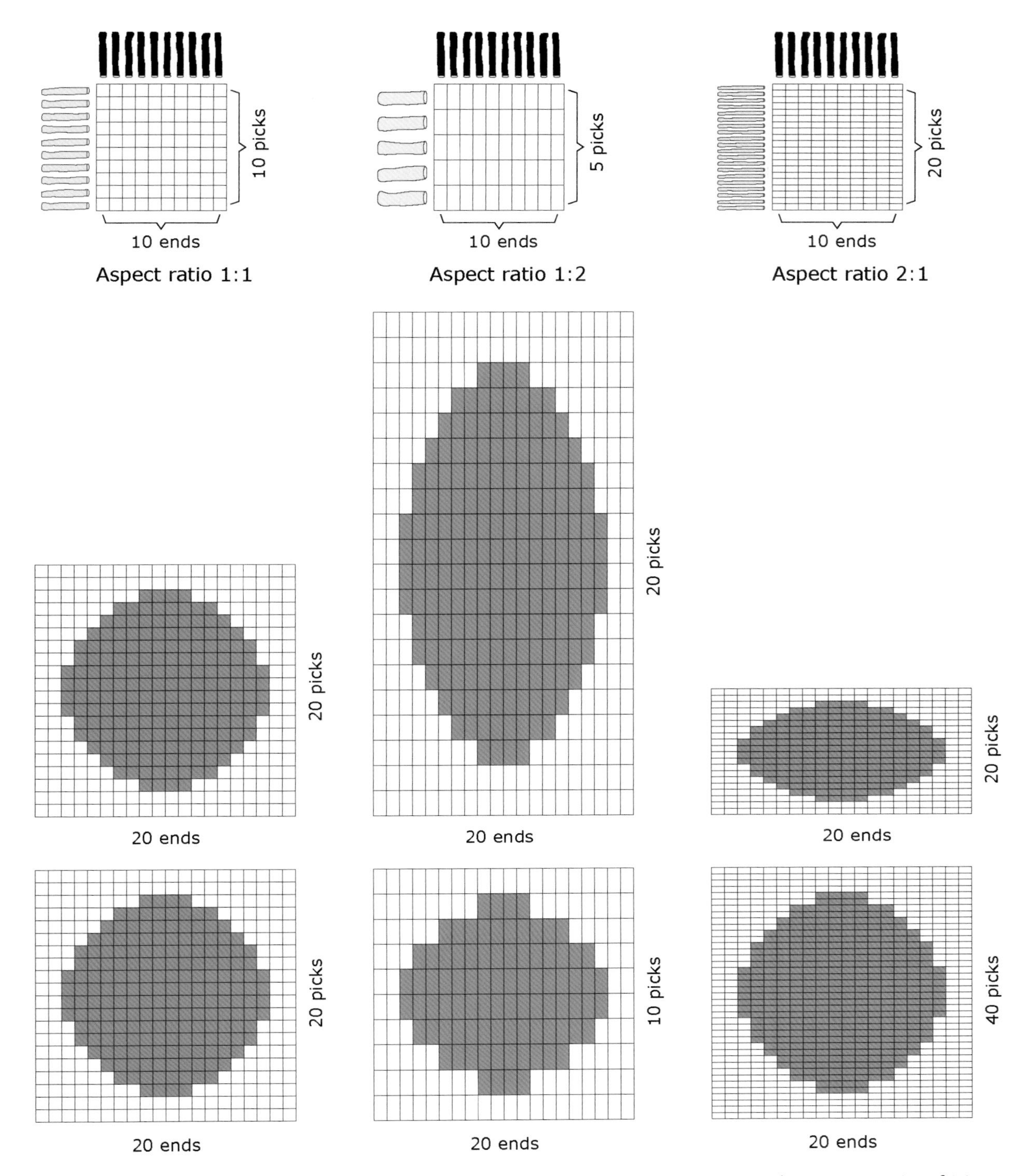

Figure 5.9. Uppermost row from left to right: *constant warp and variable weft setts produce aspect ratios of 1:1, 1:2, and 2:1.*

Figure 5.10. Pompoms *by Pradnya Korde, LFS. Handwoven damask with two self-patterning and one pattern weft.*

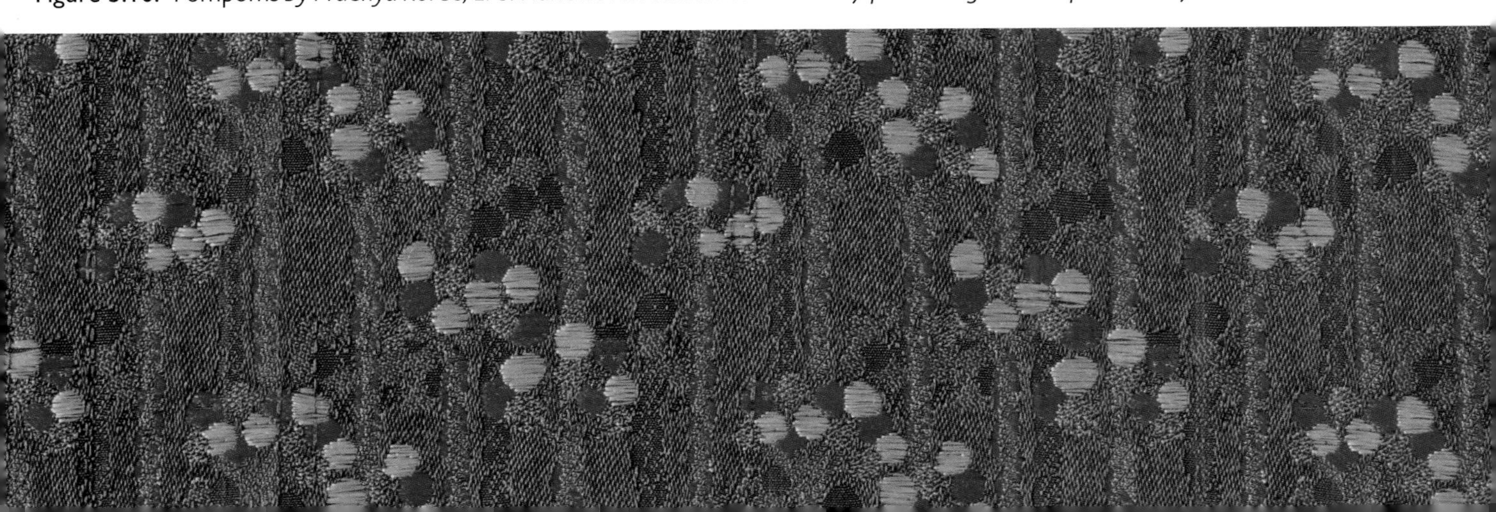

Warp line is the term used in this book for the vertical columns of a pointpaper and may be abbreviated as *wpl.* A warp line represents one or more warp ends, depending on the technique used, and the *warp step*.

Equivalent terms used in textile literature, software, and figured textile design practice are *cords* and *warp groups*. The term ***cord*** appears frequently in the handwritten record sheets found in this book and continues to be widely used in silk-weaving design.

Warp line sett is the number of warp lines relative to one linear centimeter or inch.

Weft line is the term used in this book for the horizontal rows of the pointpaper and may be abbreviated as *wfl*. One weft line may represent one or more picks, depending on the technique used, and the *weft step*.

Equivalent terms found in textile literature, software, and figured textile design practice are *passes* and *weft groups*. The term ***pass*** appears frequently in the handwritten record sheets included in this book and is widely used in silk weaving.

Weft line sett is the number of weft lines relative to one linear centimeter or inch.

Warp line value: in textiles with more than one warp series, one warp line represents two or more ends. Correct definition of this value ensures a perfect correspondence between pointpaper and finished textile.

Weft line value: in textiles having more than one weft series, one weft line represents two or more picks. Correct definition of this value ensures a perfect correspondence between pointpaper and finished textile.

In a simple textile, as shown in Figure 5.11.b, when weaving with today's looms, a warp line corresponds to one end, each weft line to one pick. Both wpl and wfl values are 1. When the same pointpaper design is used to weave a doublecloth with an equal number of ends and picks per layer, one warp line represents two warp ends—one on face, one on reverse; one weft line represents two picks—one on face, one on reverse. Both warp and weft line values are 2. (See Figure 5.11.c.)

The total number of warp lines and weft lines is equal to the visible points of interlacement, or design units, on the face, not to the total ends and picks, whereas the warp and weft line value is equal to the number of ends, or picks, each line represents.

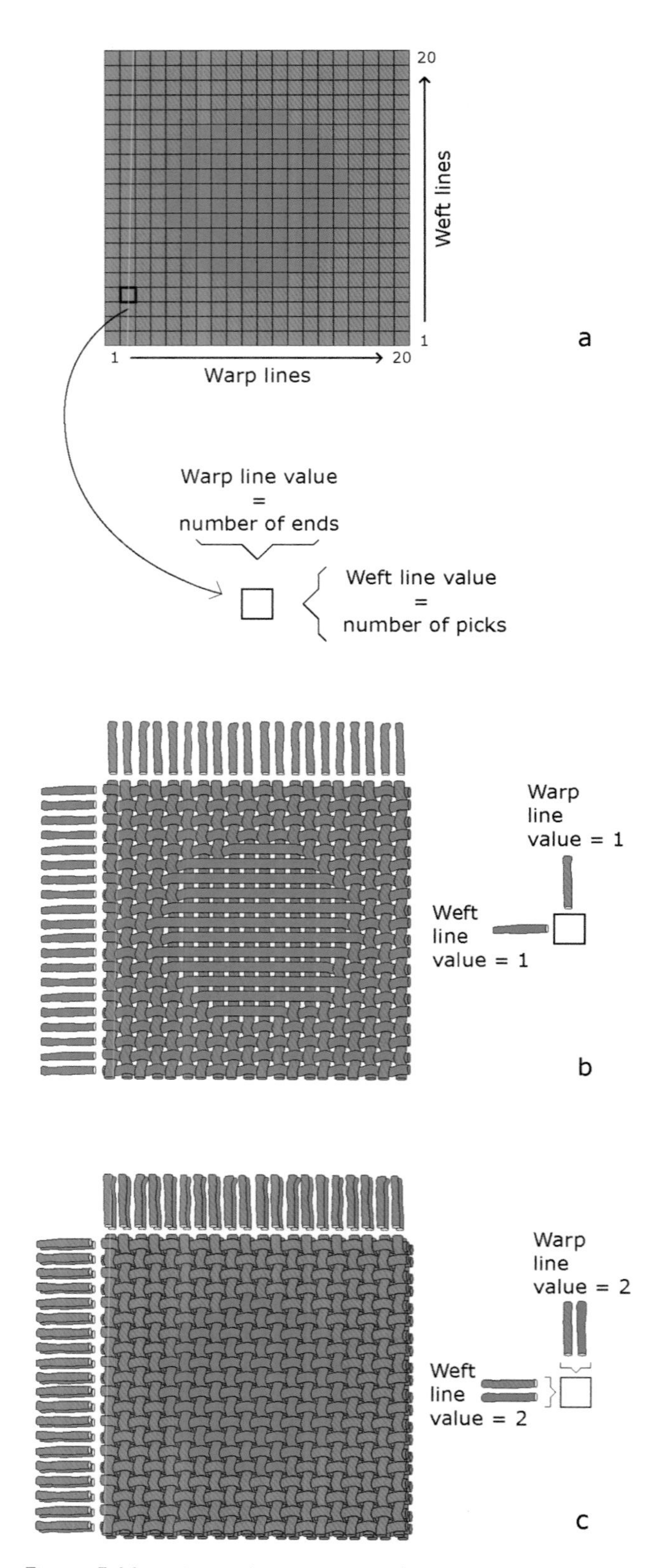

Figure 5.11. *a. In a pointpaper, warp lines represent the warp; weft lines represent the weft. One square of the pointpaper is equal to one warp line and one weft line. b. Simulation of a figured plain weave with weft float patterning; c. simulation of a figured doublecloth.*

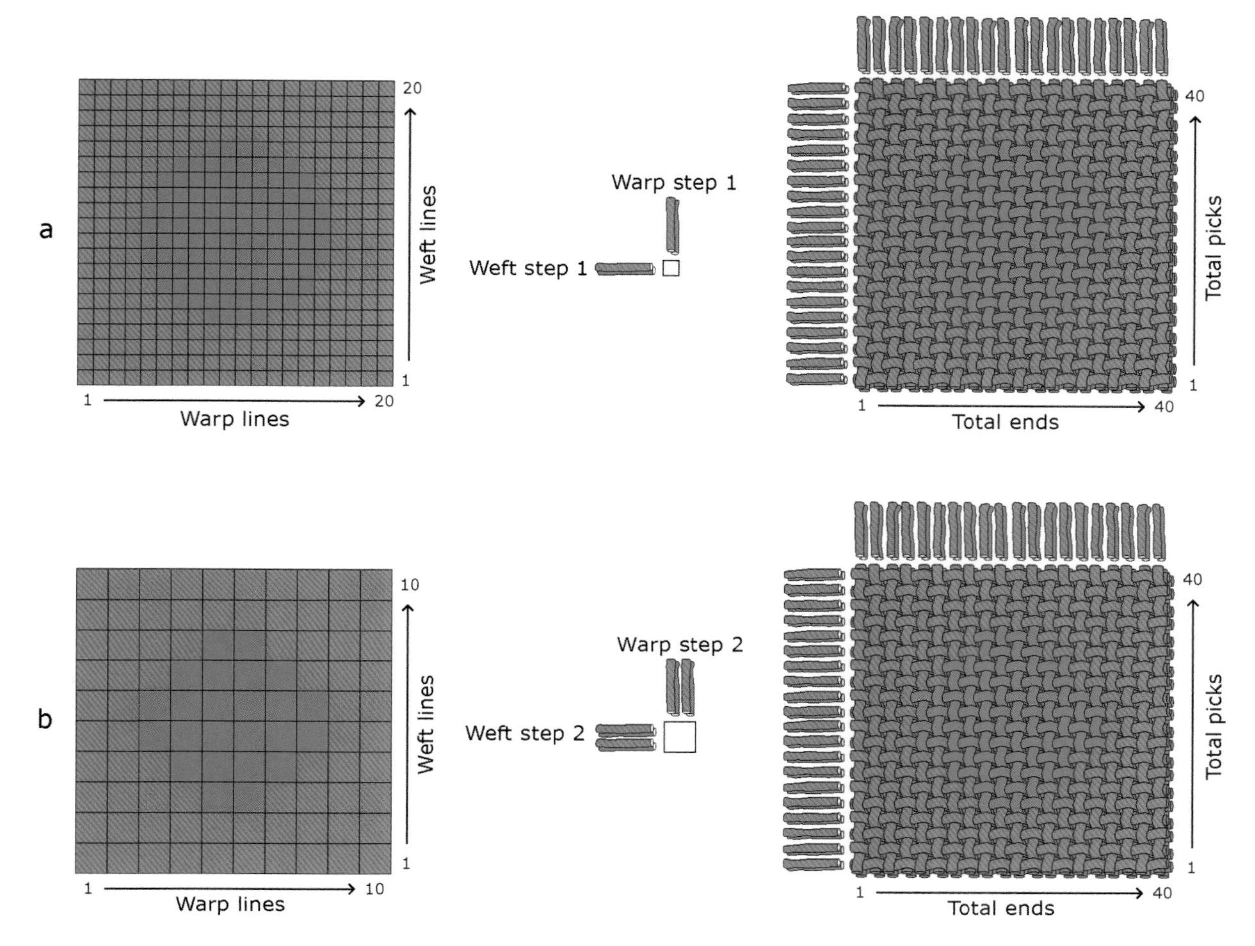

Figure 5.12. *a. Pointpaper for a doublecloth with warp and weft step of 1; b. pointpaper with warp and weft step of 2.*

Warp step is the minimum design unit in the warp direction (width), and is apparent on the textile's face where two weave effects abut.

Weft step is the minimum design unit in the weft direction (height), and can be seen at the edge of an area where two weave effects meet.

In the pointpaper and relative doublecloth in Figure 5.12.a each square represents one end and one pick on the face (and one end and one pick on the reverse). The minimum unit of exchange between the two effects, green and orange, on the face of the textile is one end and one pick. Both warp and weft steps equal 1, as is most often the case when designing for today's figuring looms. Warp and weft line values equal 2, as in Figure 5.11.c.

In a second pointpaper design b) for the same doublecloth, each square of the pointpaper represents two ends and two picks on the face, therefore warp and weft step is 2; warp and weft line value is 4.

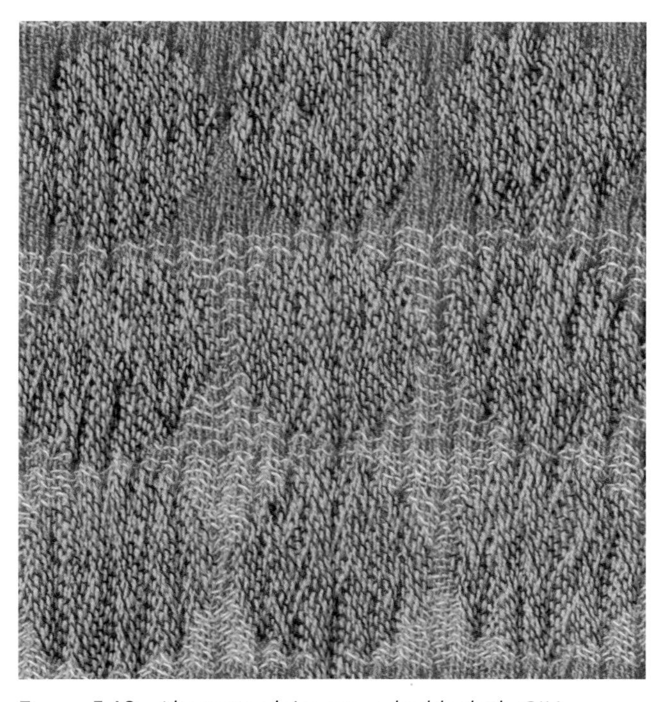

Figure 5.13. *Alternate plain weave doublecloth, CJH.*

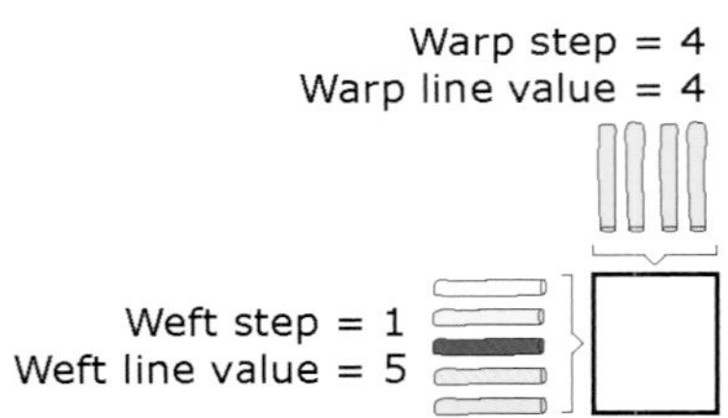

Figure 5.14. Above: *Detail of brocaded plain weave; the minimum pattern unit is composed of four warp ends and five picks.* Facing page: Santa Maria del Fiore *shown at real size. Courtesy of LFAS.*

Warp step and weft step are often noticeable in textiles woven on handlooms set up with separate ground and figuring harness, the second equipped with stepped mails (the heddle of a figuring harness), that raise multiple ends as a group to create the patterning, thereby widening the design capacity of a given loom. (See Figure 6.8 in Chapter 6.)

In the detail of *Santa Maria del Fiore*, a brocade with one warp series, one ground weft and multiple brocading wefts, a *warp step* can be seen where the ground weave abuts brocaded areas. The minimum unit of exchange between weave effects in the warp direction is four ends, and therefore warp step is 4. As there is only one warp series, and all of these are visible on the face of the cloth, warp line value is also 4, that is, one point or square of the pointpaper represents four ends of the warp.

Observation of the edge of the brocaded areas in the weft direction shows a minimum pattern unit of 1. As many as five weft series (one ground and four colors) compose one line of patterning in a given motif. The weft step value is 1; weft line value is 5.

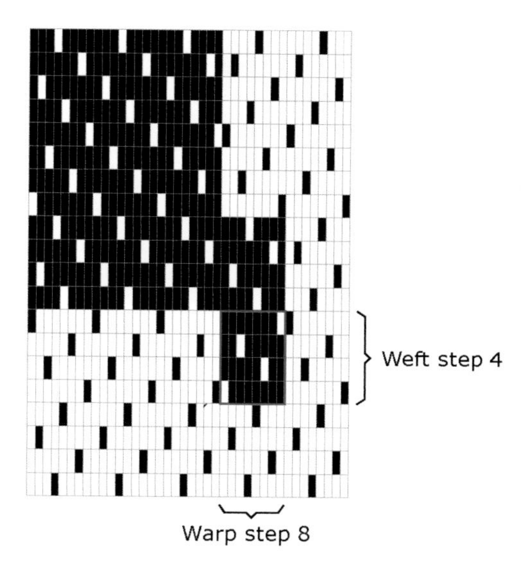

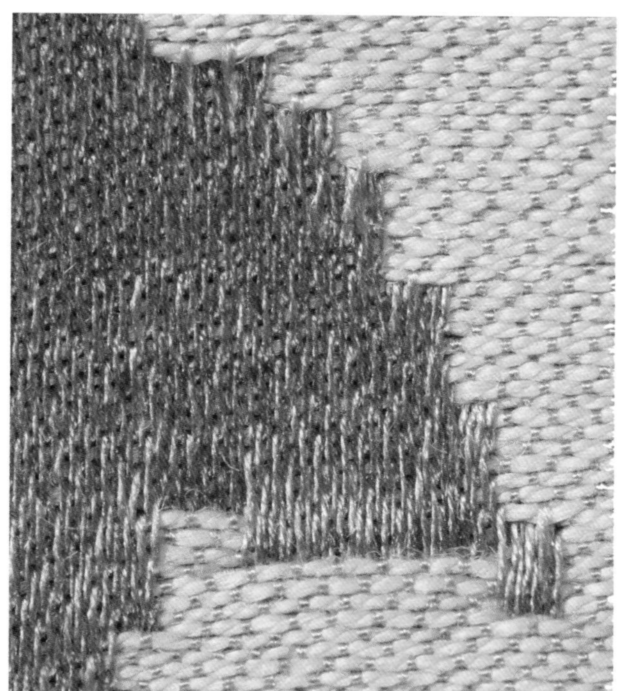

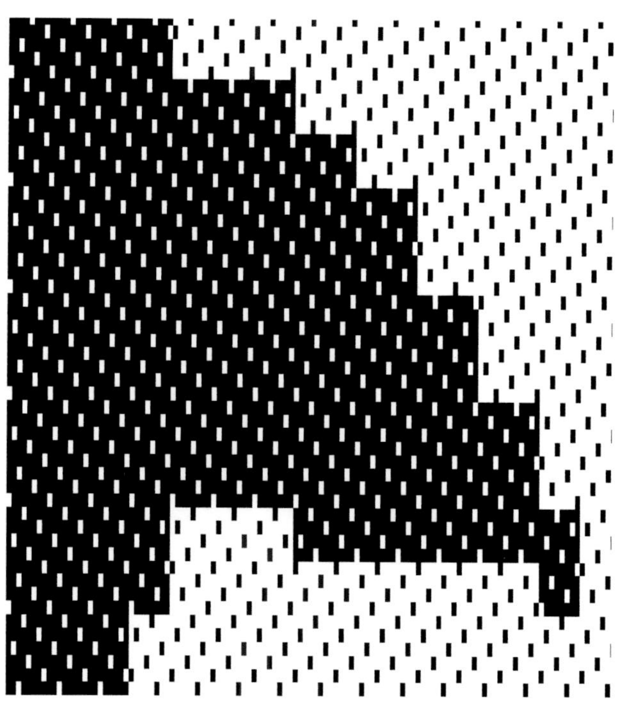

Figure 5.15. Facing page: *Damask with a warp step of eight and a weft step of four,* Graffiti *by Katie Isaia, LFS;* This page, top: *Detail from the card used to weave* Graffiti *showing the minimum unit of exchange between effects—warp step of eight and a weft step of four.* This page, bottom: *Detail of sample and card showing the "stepped" outline between two weave effects.*

Both warp and weft step may be used with today's single hook/end patterning looms to create a "stepped" or pixelated effect at the edges of design areas, or to simulate the look of an older textile. In *Graffiti*, a damask with one warp and weft series, the designer enlarged her original design and produced a stepped outline at the edge between weave effects by assigning eight ends to each warp line and four picks to each weft line of the pointpaper. Warp step is equal to eight; weft step is equal to four.

Effect: The term effect, or ***weave effect***, is employed for each distinct weave structure, simple or compound, that appears on the surface of a figured textile. Each effect is drawn on the pointpaper using a distinct color. Colors used to draw effects do not represent the final appearance of a weave or materials; bright, easy to see colors may be chosen to facilitate the designer's work. Depending on an effect's role within the design, it may be called a ground effect or pattern effect. Synonyms for pattern effect are patterning or figuring effect.

Figure 5.16. Left: *Two repeats in width and height of the sketch for* Sticks. Below: *The pointpaper design, shown in the* foreground, *is drawn in three colors that represent three distinct weave effects.* Background: *warp-faced broken twill ground with two self-patterning wefts.* Sticks *by* Tuulia *Lampinen, LFS.*

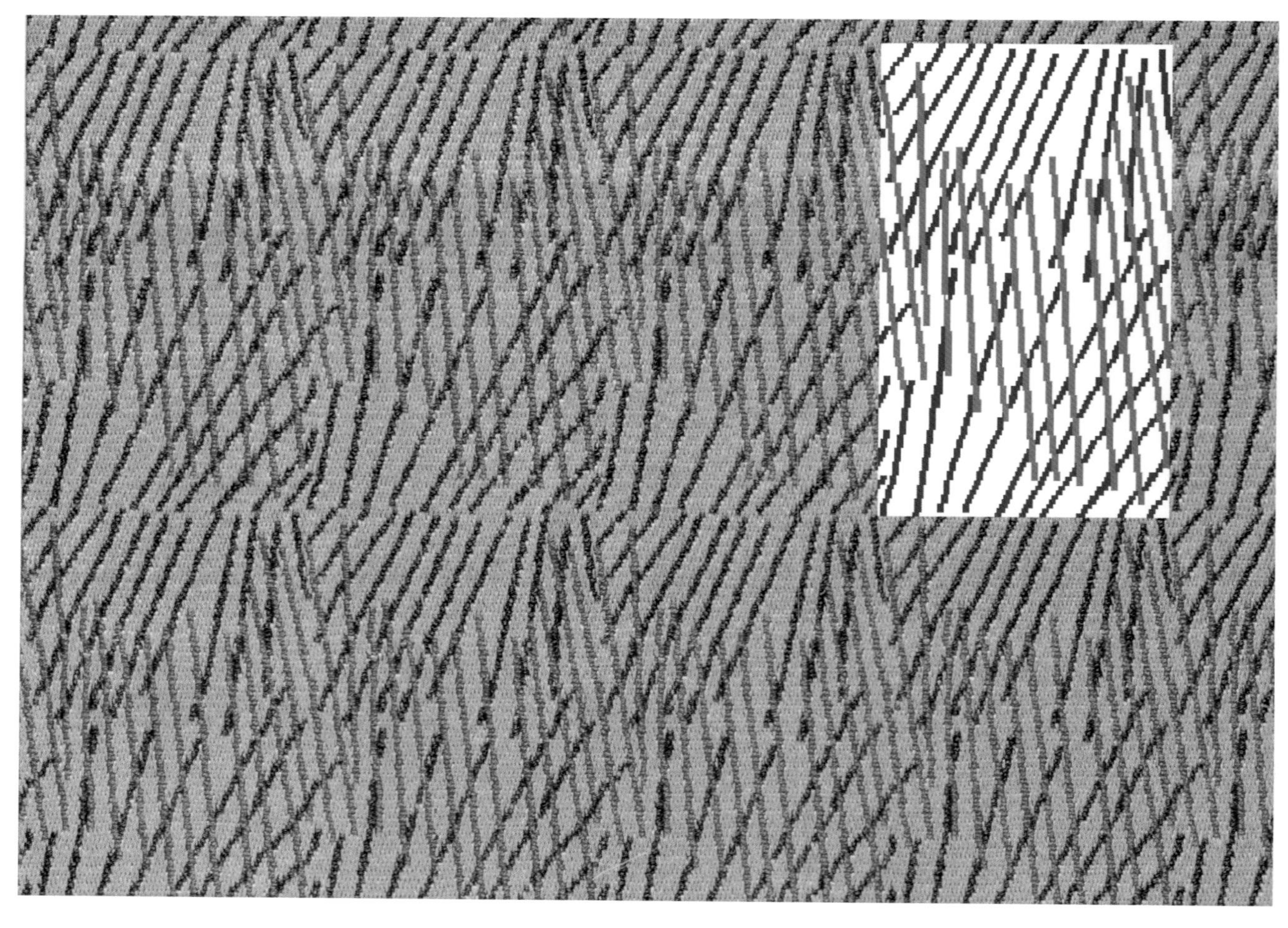

READING NOTE : FACE DOWN

1 cord of graphpaper : 1 hook
1 pass of graphpaper : 2 cards

	ground	EF 1	EF 2
O	TABBY, TWILL OR SATIN PER PASS	ALL UP	——
O		——	ALL UP

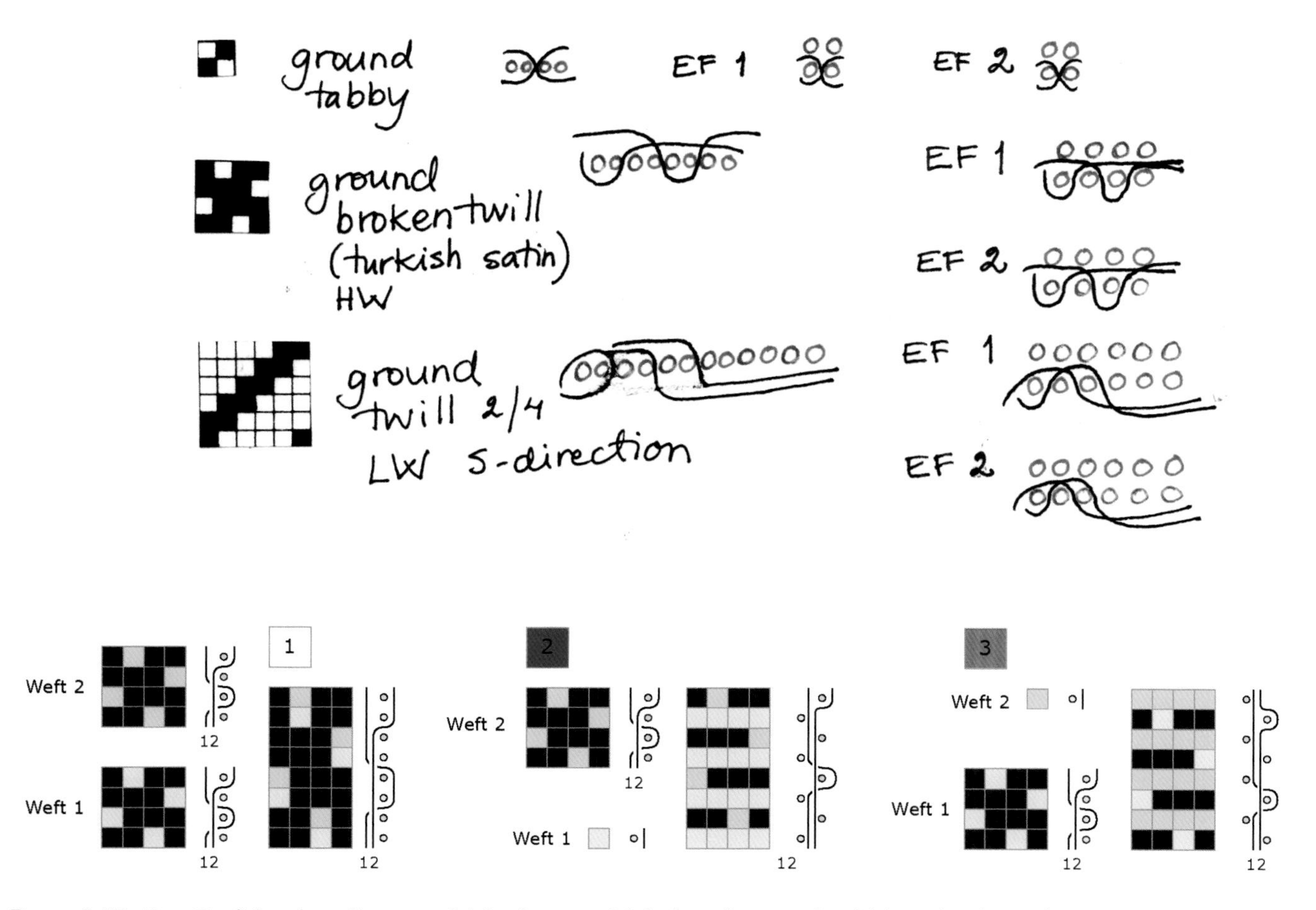

Figure 5.17. Top: *Traditional reading note,* Sticks. Bottom: *Digital reading note in which each color in the pointpaper is associated to a weave structure drafted on squared paper.*

Reading note: whether in the graphic or verbal form, the reading note establishes a correspondence between each color and weave effect in a figured textile. The hand drafted reading note in Figure 5.17 lists three alternative ground weave structures—tabby, broken 4-end twill, and 2/4 Z twill, all woven with doubled picks—for three versions of the design in Figure 5.16.

Simulation: Those using textile design software may define all the drafts that compose the textile—pointpaper, reading note, warp/weft sequence and sett, colors, and materials—and then simulate a "virtual" textile on a screen or as a printout. This preview of the textile allows the artist/designer to evaluate the effectiveness of imagery, weave choices, color, materials, and so forth, and make eventual corrections before weaving a physical textile.

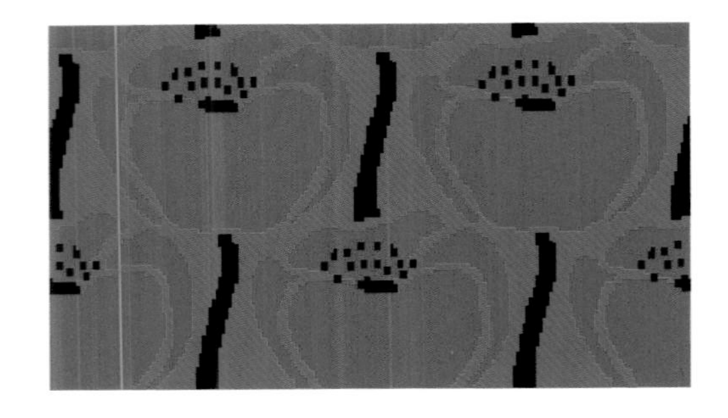

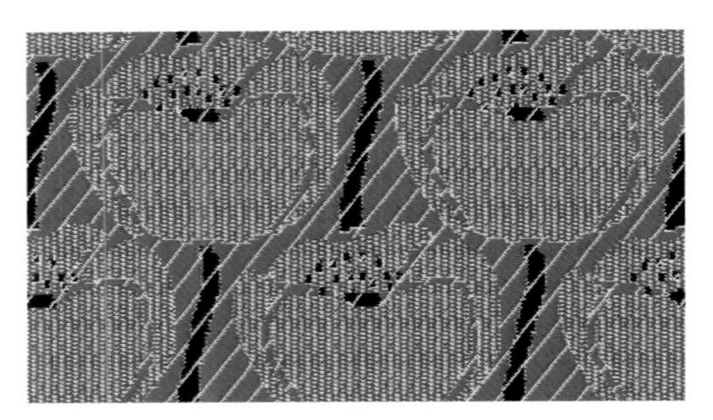

Figure 5.18. Above: *Digital pointpaper design and simulation for* Poppies *by Lut Verrelst, LFS.* Below: *Photograph of the finished sample with unpainted and painted warp.*

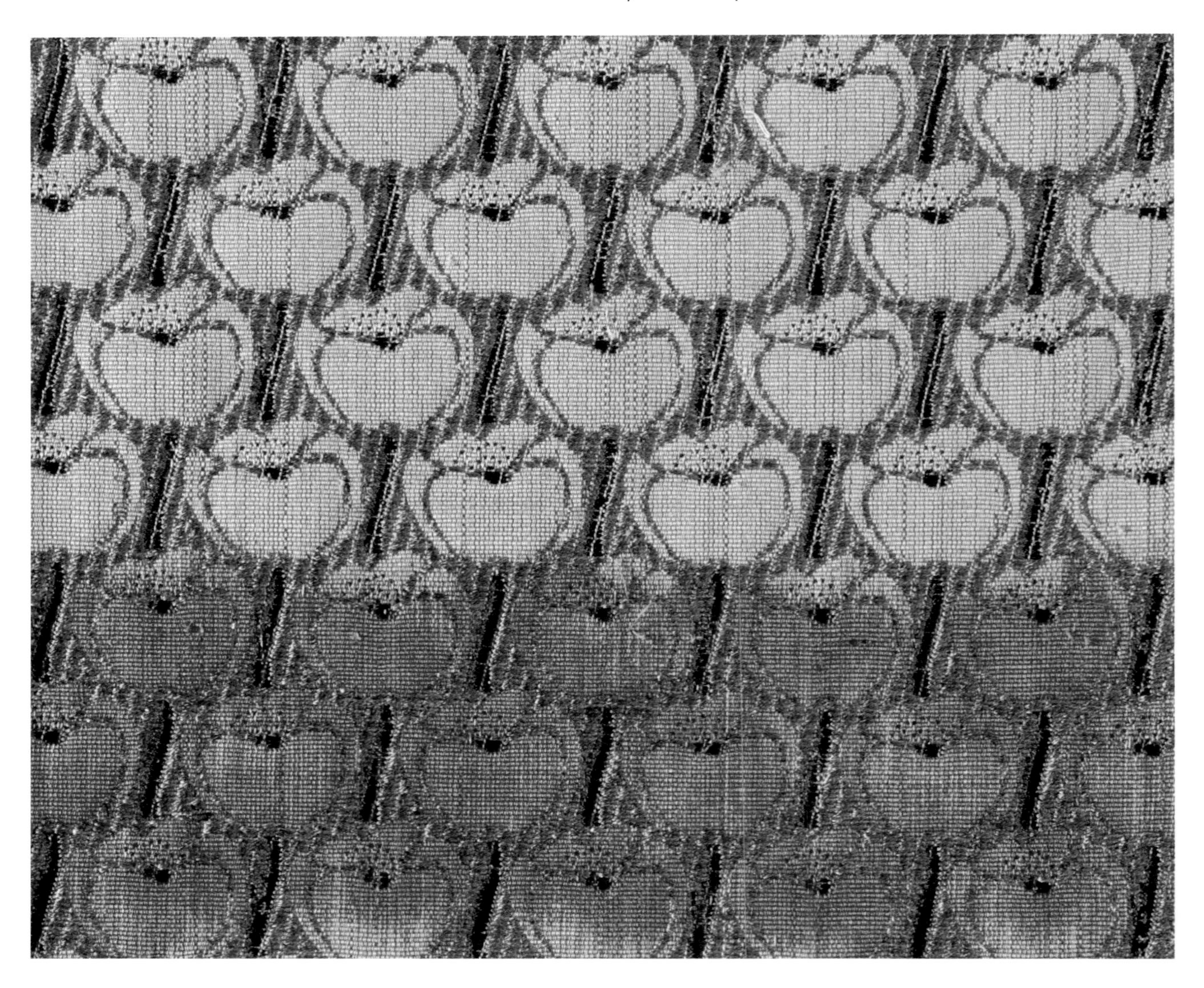

The *Card*: whether physical or digital, the card is a draft that records the position, raised or stationary (action vs. no action, hole vs. no hole, black/filled square vs. white/empty square), of every end relative to every pick of a figured textile's repeat. Each horizontal row of the card represents one pick; each vertical column represents one warp end.

During weaving the card is transmitted, one pick at a time, by the loom's figuring device to the warp.

Figure 5.19. Above: *Card for* Poppies. Below: *Detail of finished sample.*

6 Figuring techniques

A selection of figuring techniques, from simple to complex, is presented in the following pages together with photographs that illustrate the aesthetic potential of each class. The accompanying drafts show possible solutions to the challenges posed by the single techniques. Self-patterning, weft-patterned, warp-patterned, multiple cloths, and lampas are just a few of the myriad of weave-patterning techniques used in craft, art, and industrial weaving. With today's looms, hybrid use of these and other techniques is common; many textiles could be defined as belonging to several classes at once. Slow looms, fast looms, old and new technologies can be used together to produce hand or industrially woven figured cloth that follows standard canons, or breaks down the boundaries between art, decorative, and functional textiles.

One warp, one weft: self-patterning textiles

Self-patterned textiles employ ground warp and weft series to create both the ground weaves and patterning. Even if warp and weft are identical in sett and materials, minimal changes in weave structure create a contrast on the surface of the cloth.

Figured Plain Weave

The simplest of self-patterned textiles is figured plain weave patterned by warp or weft floats.

Figure 6.1. *Plain, warp up, warp down.*

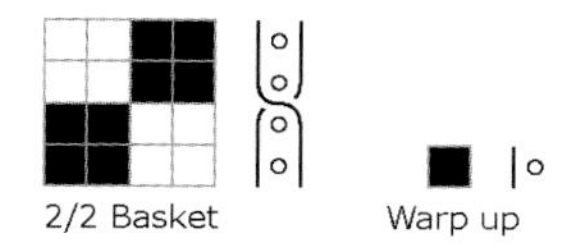

Figure 6.2. *Drafts for figured 2/2 basket with warp floats, shown on facing page, CJH.*

The textile in Figure 6.3, *Vanessa I,* is woven with the best-known self-patterning technique: damask.

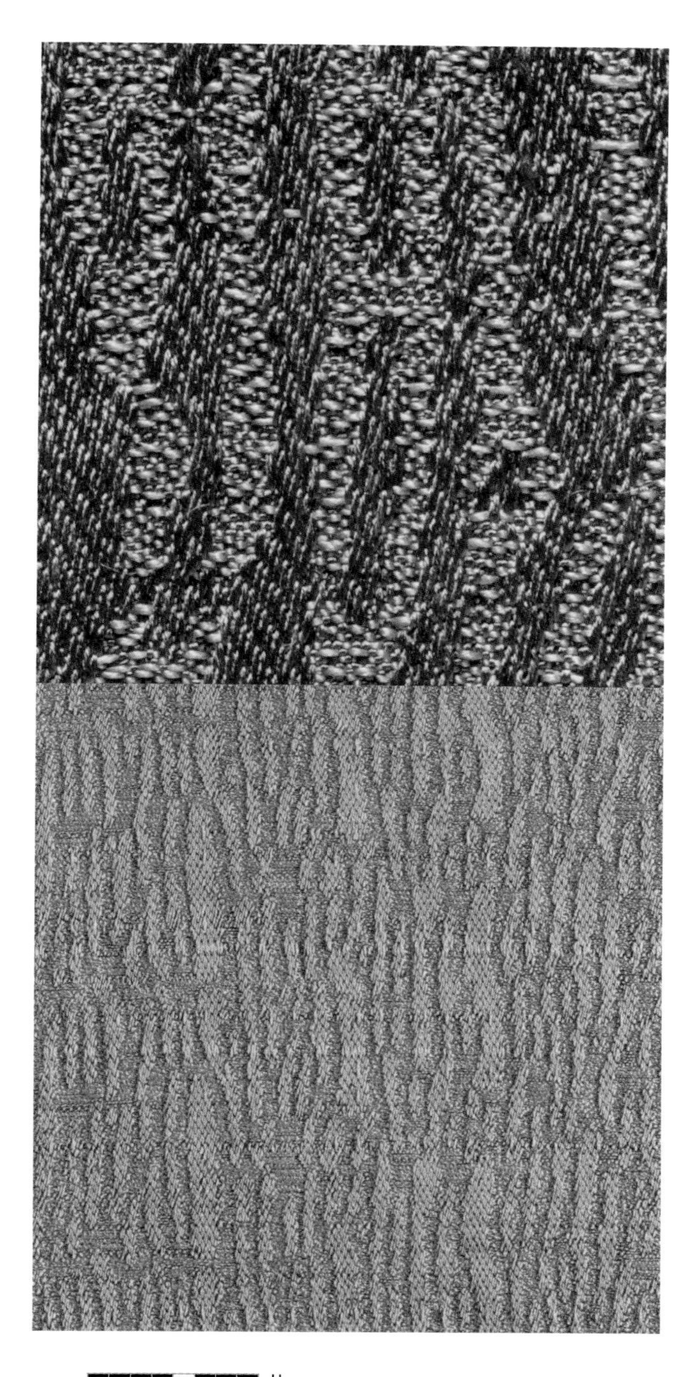

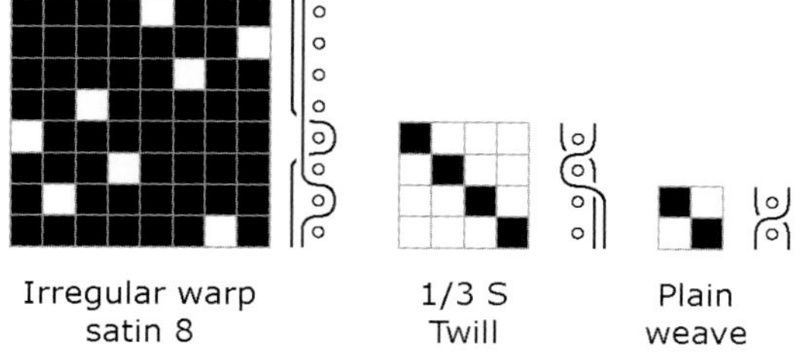

Figure 6.3. *Three weaves of the damask* above. *Detail and 1:1 photograph,* Vanessa I, *by Vanessa Golestaneh, LFS.*

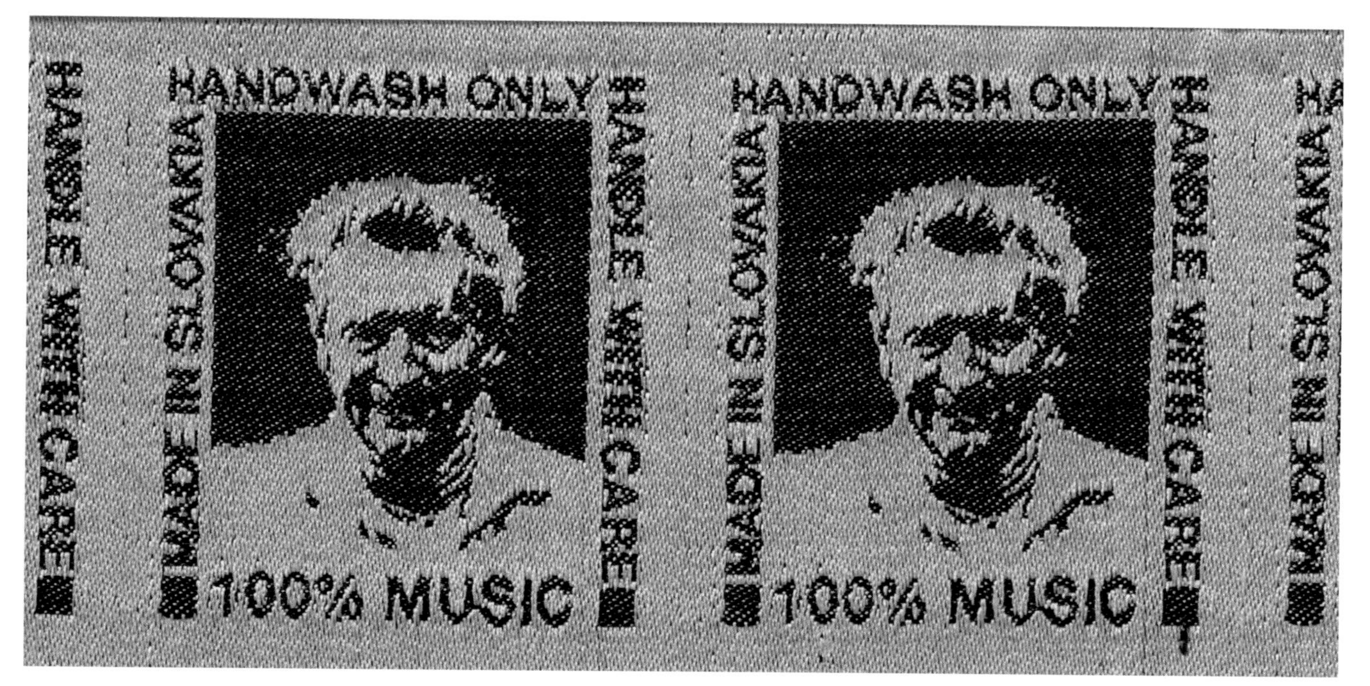

Figure 6.5. *Classic damask woven in warp and weft satin 8,* Handwash Only, *by Lut Verrelst, LFS.*

Damask

A technique that first appeared in ancient China during the Tang dynasty[1] (618–907 C.E.), *damask* is woven with one warp and weft series and simple weave structures. In its most commonly known form, damask is woven with the warp and weft-faced version of the same weave structure, such as warp and weft satin 8. The resulting changes in light reflection create patterning on the textile's surface.

Warp and weft-faced weaves may be combined with balanced weaves, weaves with larger or smaller repeats, and different weave classes, as shown in the table of traditional damasks with two weave effects in Figure 6.6.

Name	Ground weave		Pattern weave	
Damask Satin 5	Warp satin 5		Weft satin 5	
Damask Gros de Tours	Warp satin 8		2/2 Basket (Gros de Tours 2 ends)	
Lyon Damask	Warp satin 8		1/3 S Twill	
India Damask	Warp satin 8		Weft satin 8	
Lustrine Damask	Warp satin 5		1/1/1/2 Z Twill with doubled picks	
Louisine Damask	Warp satin 8		2/1 Rib (Louisine) on alternate picks	

Figure 6.4. Facing page: *Daffodil-patterned table linen woven in warp and weft satin 5, CJH.*

Figure. 6.6. Above: *Table of traditional damasks with two weave effects.*

The Traditional Damask Loom

Today it is increasingly common to design and weave with equipment that allows single-end control and a selvage-to-selvage repeat. We take for granted the ease with which designs are transmitted to the patterning device of a loom. Before computer technology, ingenious loom setups were devised to reduce the amount of time and labor necessary to transmit patterning information to the warp.

The silk damask shown in Figure 6.7 has a low sett at circa 65 ends per centimeter (165 ends per inch) and a narrow repeat eight centimeters (3.14 inches) wide; woven today, it requires a loom setup with 520 independently controlled warp ends. In the past damask was woven with a rising and a sinking harness to produce the warp and weft-faced weaves typical of this technique, while a separate figuring harness, mounted with *decked* or *stepped* mails (a heddle with multiple holes or loops), raised or lowered groups of ends to create the patterning. Figure 6.8 illustrates how each end was threaded separately on both a rising and a sinking harness that wove the ground; multiple ends were then threaded in groups through the mails of the figuring harness. As weaving proceeded, the decked mails (and groups of ends) were raised by pulley cords to create the design. With the device illustrated here, a 520-end repeat could be woven with 104 pulley cords, each of which controlled 5 ends. In damasks woven with these traditional setups, one can see the characteristic "step" created by a figuring harness with decked mails.

Figure 6.7. *Detail and photograph* (above and facing page) *of classic silk damask in satin 5 on striped warp, LFCHT.*

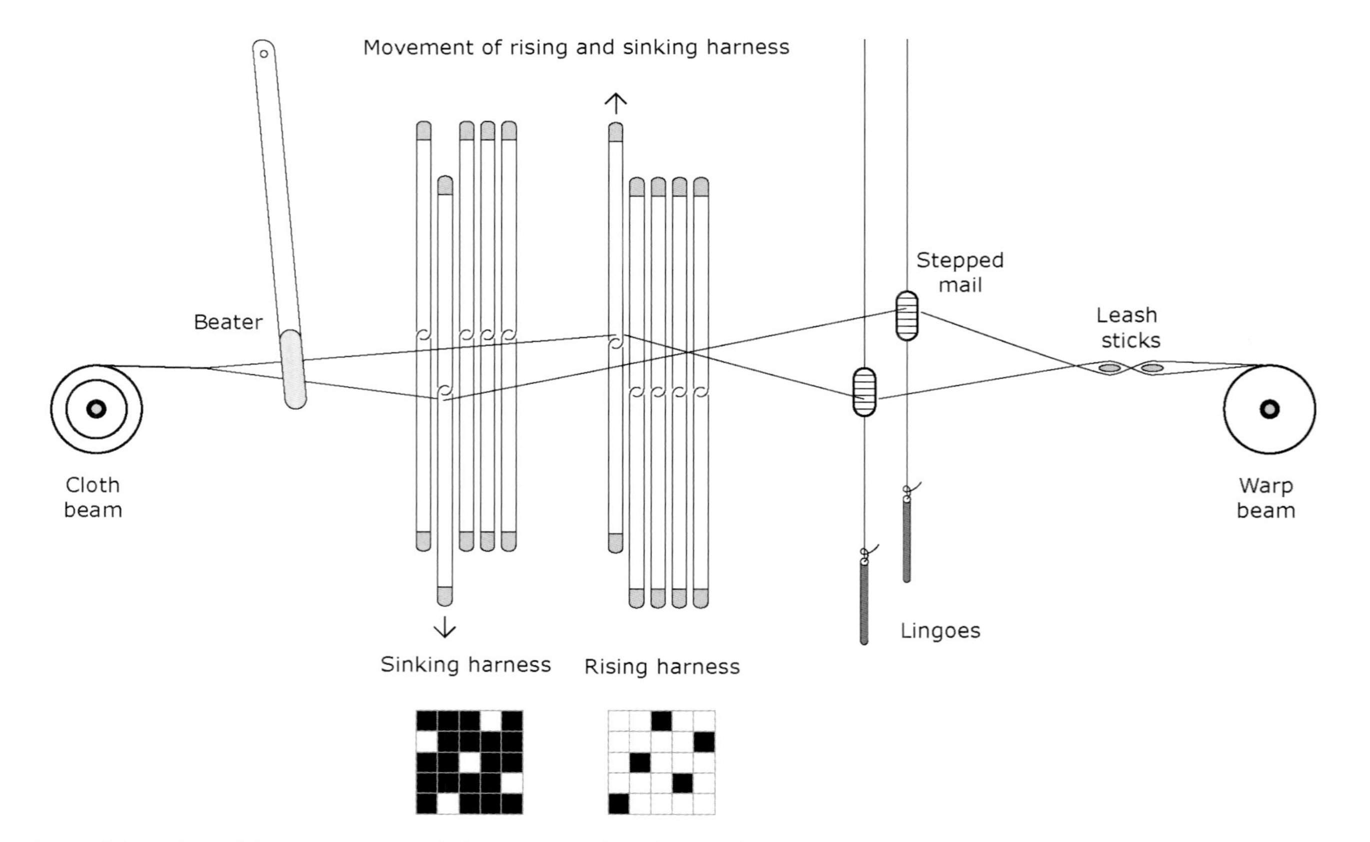

Figure 6.8. *A damask harness composed of rising and sinking harness for the ground and a figuring harness to create the pattern.*

When designing damask, the first step is to identify the kind and level of contrast required. The designer thinks in terms of light reflection, as seen in the classic white-on-white damask in Figure 6.4 and other tone-on-tone damasks found throughout this book. In addition to light reflection, contrast may be produced by differences of sett, material, color, warp or weft design, or weave structure.

Structurally, *Triangles* is a classic tone-on-tone damask, woven with a ground in warp satin 8 and patterning in weft satin 8. The many horizontal lines of the geometrical design create what in most textiles would be a defect: a series of raised bars across the width of the textile, where the densely sett, silk warp face exchanges with woolen weft-faced patterning. A combination of design, different warp and weft setts, an elastic silk organzine warp, and a fine, spongy wool weft generates an emphatic waffled effect over the surface of the textile.

Onions was woven on the same silk organzine warp using the same weaves, but with a cotton weft that produces ridging along the exchange between warp and weft surfaces. Note that the raised effect is absent where the design's lines are vertical, reappears as the lines follow a diagonal path, and is more pronounced where the exchange between warp and weft faces falls horizontally. Stronger or subtler colors of a sampling warp throw the design into greater or lesser relief; both tone-on-tone and high-contrast colorways are effective.

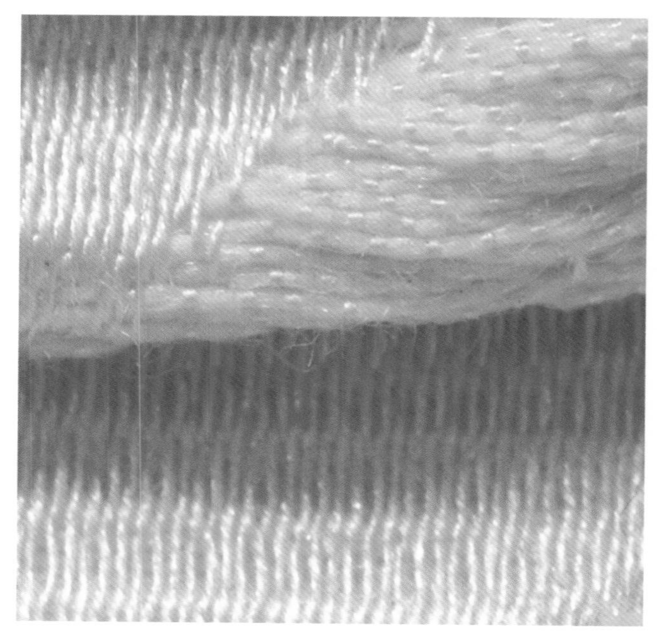

Figure 6.9. Left: Triangles *by Annette Schryen, LFS;* Above: *detail.*

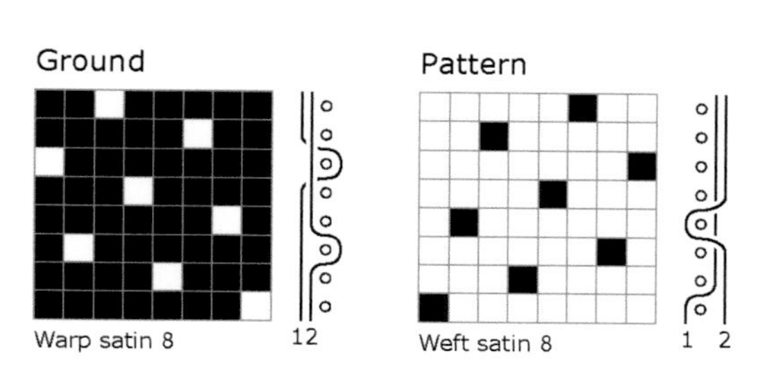

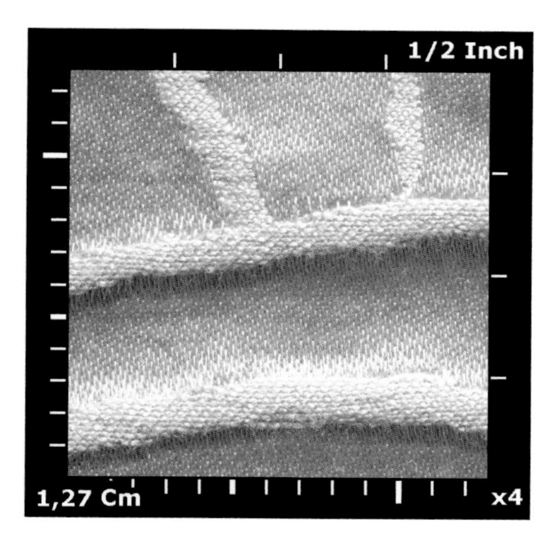

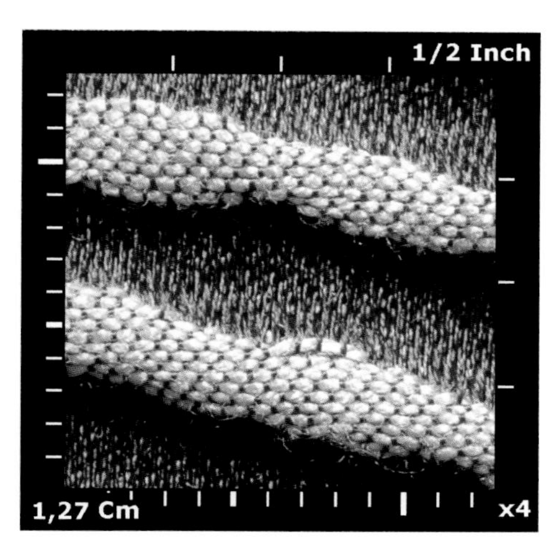

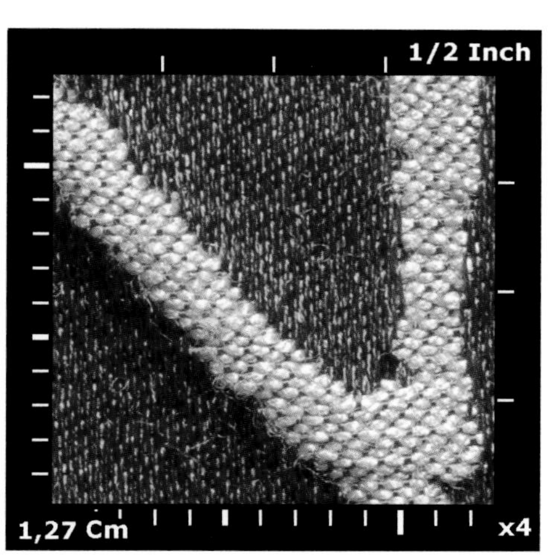

Figure 6.10. Above: *weaves; details;* Right: *photograph of* Onions *by Hans Thomsson, LFS.*

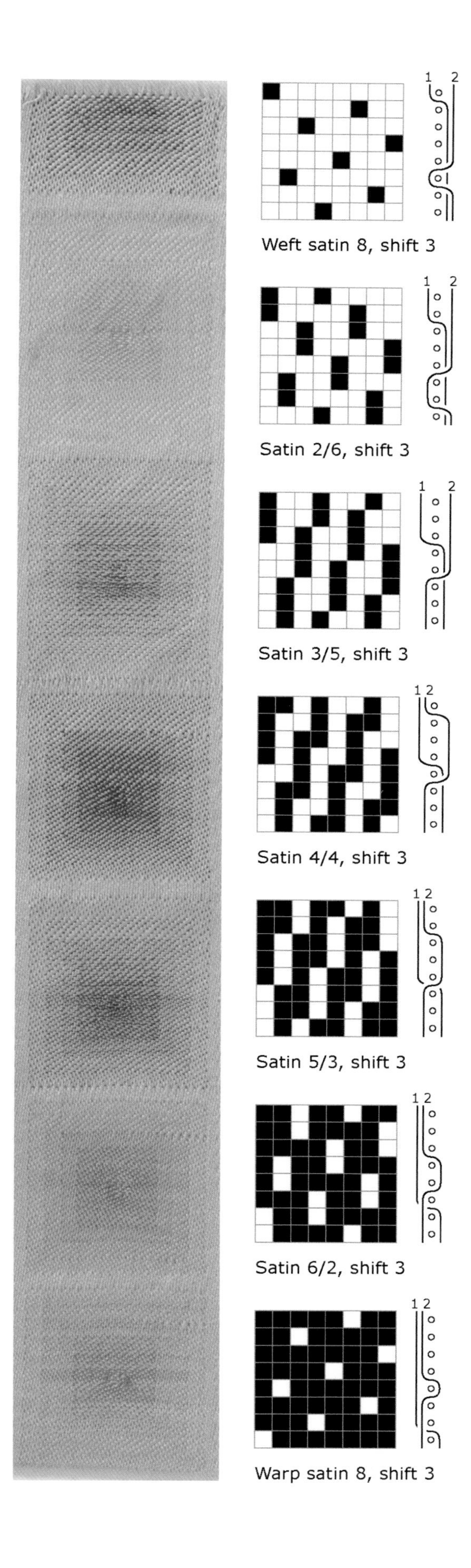

When woven with warp and weft in contrasting colors, damask is a technique that lends itself to the creation of shaded effects. By using weaves of the same class and repeat size that progress from warp to weft surface or the reverse, gradual transitions can be generated that give a sense of dimension to the textile. Twills and satins are particularly suited to this use, as the orderly addition or subtraction of rising ends from a drawdown produces weaves that join seamlessly and that do not generate added take-up to any one end or group of warp ends.

In the drafts for industrially woven *Damask Squares* in Figure 6.11, note how seven shaded satins progress from weft to warp face by the addition of binding points in a vertical direction. Equivalent series of shaded satins and twills can be made by systematically adding binding points in the horizontal direction, a choice the same designer adopted for a smaller, handwoven shaded satin damask, *Shaded Squares,* on this page.

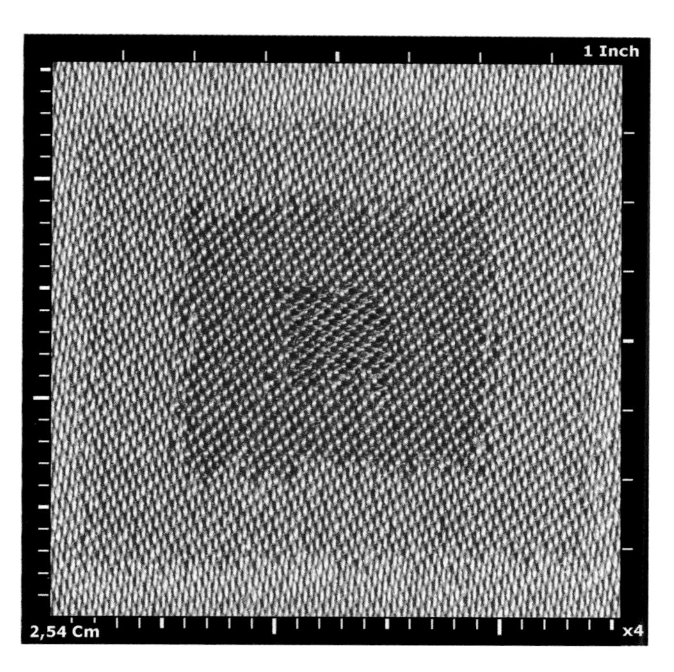

Figure 6.11. Facing page: *industrially woven* Damask Squares. This page, from left to right: *handwoven* Shaded Squares, *seven drafts for the seven weave effects used in* Damask Squares, *detail of* Damask Squares. *All samples by Helena Loermans, LFS.*

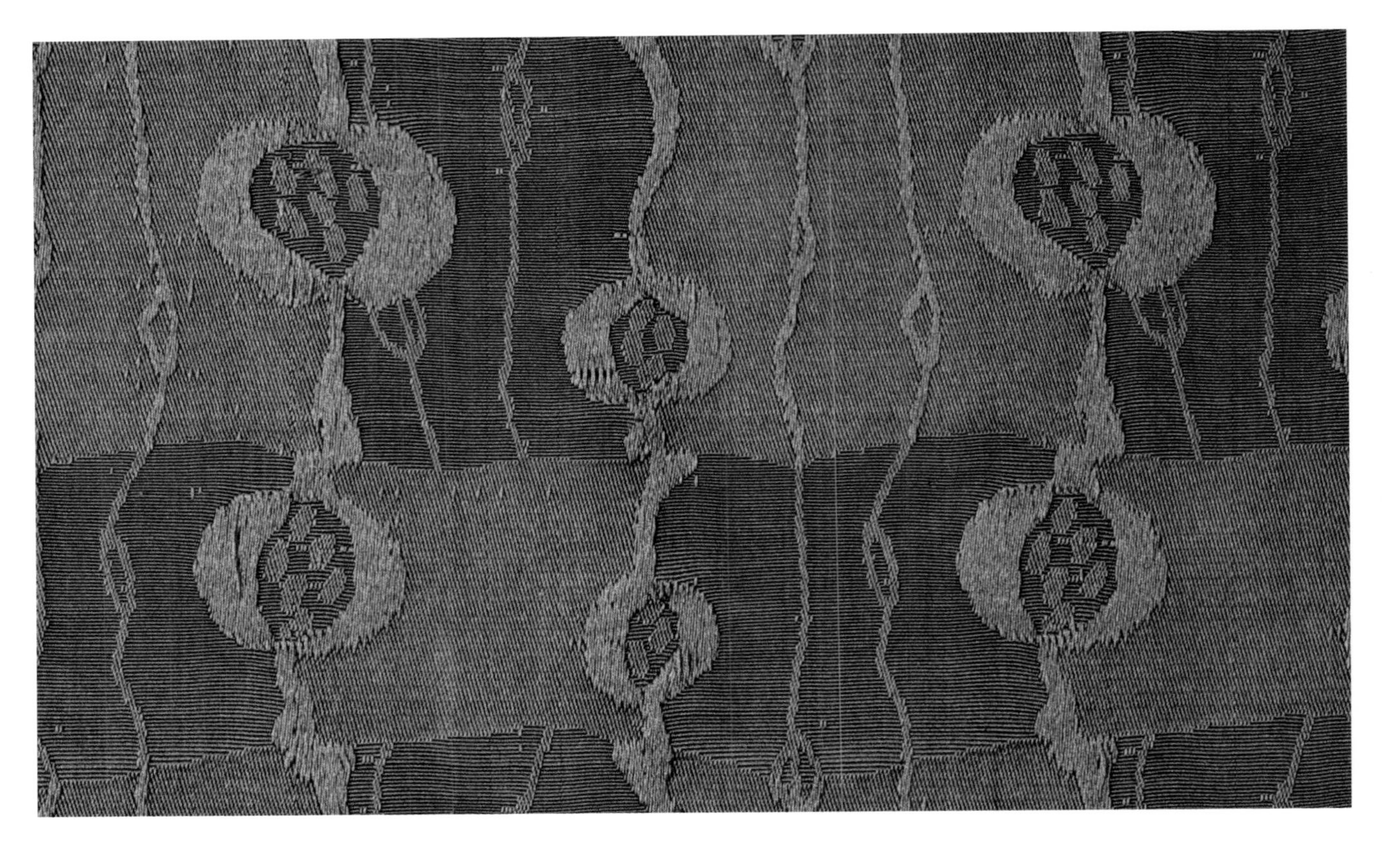

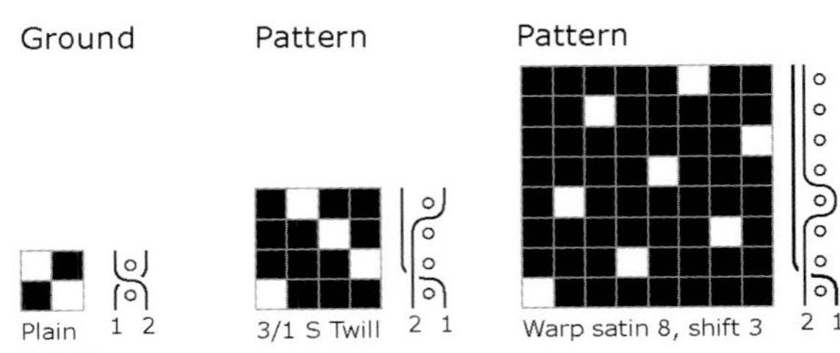

Figure 6.12. Top: *Weaves for* Pomegranate, *silk damask with three effects by Päivi Fernström, LFS.*

If there are more than two weave effects in a design, the maker considers how to "spend" warp and weft potential for contrast. With the textile's design and materials in mind, the maker selects weaves that present different percentages of warp and weft, as well a range of textures: the glossy, smooth surface offered by satins; the pebbly, matt surface of smaller weaves such as plain, 2/1 or 1/2 rib, 2/2 basket, or the distinct graphic line created by the twill group. The designer might choose the glossiest surface for motifs in the foreground, while the second most apparent surface could be assigned to the background, leaving a blended effect for the middle ground.

Each of the three structures used to weave *Pomegranate* is from a different weave class and has a different repeat size: satin / 8, twill / 4, tabby / 2. Each effect presents different percentages of warp and weft on the cloth's surface. Note that the high take-up of plain weave, combined with the lower take-up of satin 8 and placement of weave effects in the design, generates differences in warp tension. By substituting plain weave with an extended tabby, tension differences would be lessened without altering the weave class of the ground effect.

In the silk and cotton damask *Grass,* three 8-end weaves are used to create gradations from warp to weft-faced surfaces: warp satin 8, 2/6 shaded satin, and 1/7 twill. Two different weave groups are present: satins and twills. Warp and weft percentages vary in each effect, while weave repeat size and take-up remain constant.

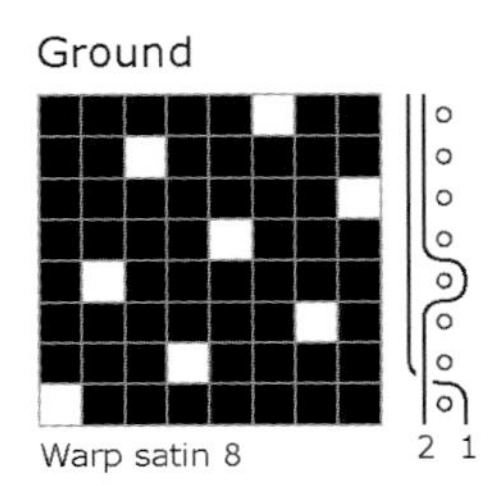

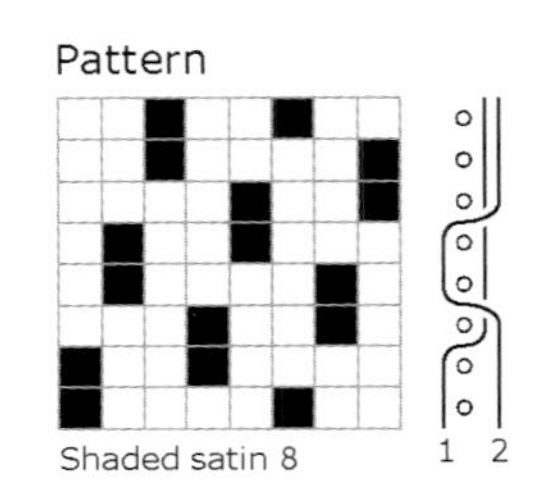

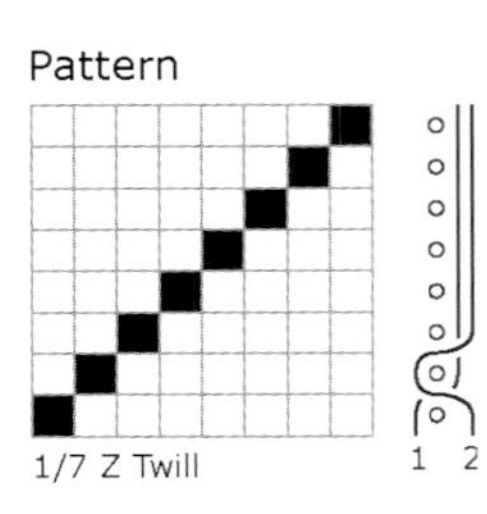

Figure 6.13. Left and facing page: Grass *by Carmel d'Ambrosio, LFS, woven on a striped sampling warp; weaves.*

Weft patterning

Self-patterning Wefts

There are textiles that are woven without compromise, in which time and materials are employed with no regard to cost. Slow weaving with rare materials may be the exercise of an art, a discipline; the production of a unique textile may be ceremonious or have significance, but textile economies are also beautiful as exercises in textile logic and expertise; they help us to use our materials, equipment, and time to best advantage.

Every pick added to a simple textile (one warp and weft series) adds material, cost, weight, and weaving time. The use of ground picks to create weft patterning is one of the great economies of figured weaving.

Gondola's ground is woven as a simple weave, a 1/2 rib, using one warp and one weft series. Where the figuring occurs, the weft is divided into two series of picks, each with a distinct function: the odd picks weave below to create a plain weave *bed,* while the even picks float above to create the figuring, bound as required by binding points drawn on the pointpaper. These binding points correspond to one raised end used to tack down overlong floats. This technique, called self-patterning, obviates the need for an extra pattern weft. Note that for purposes of take-up, or warp shortening, the amount of interlacement is constant in both ground and patterning effects, with one exchange of the warp's position from face to back occurring every four picks.

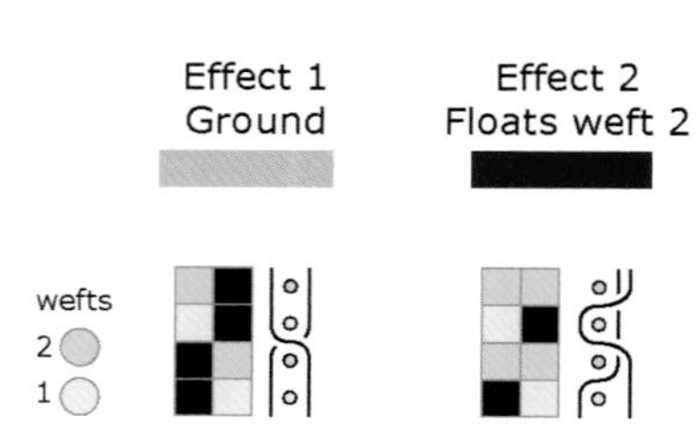

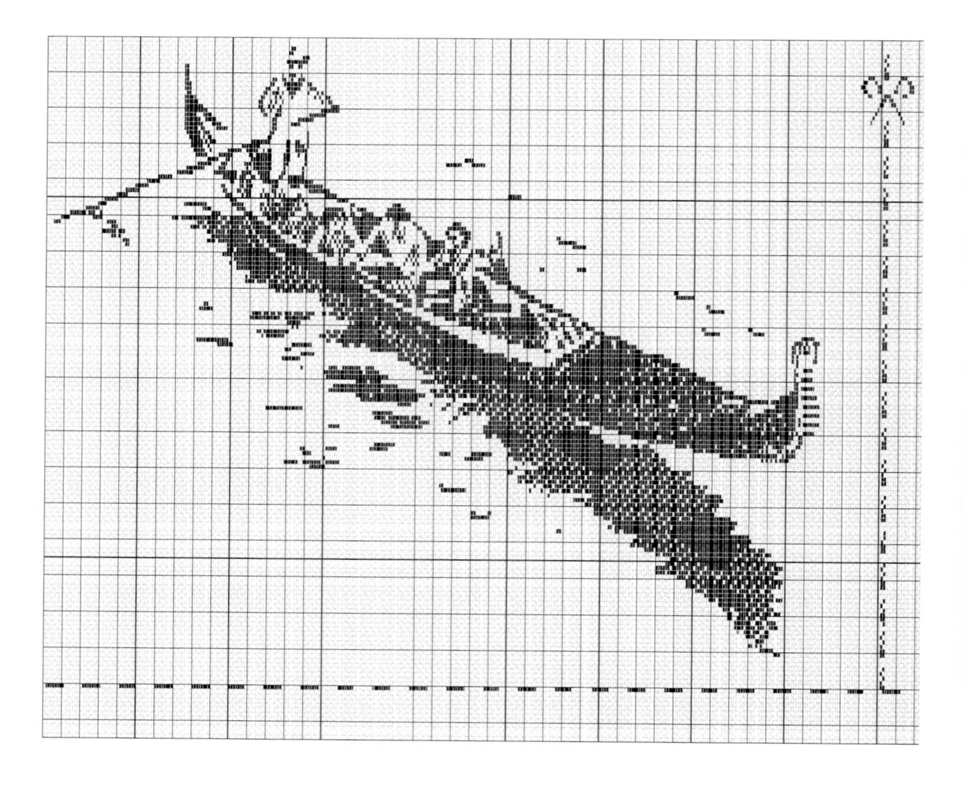

Figure 6.14. Gondola, *figured silk 1/2 rib ground with self-patterning weft effect in silver by Tina Moor, LFS; weaves; pointpaper design.*

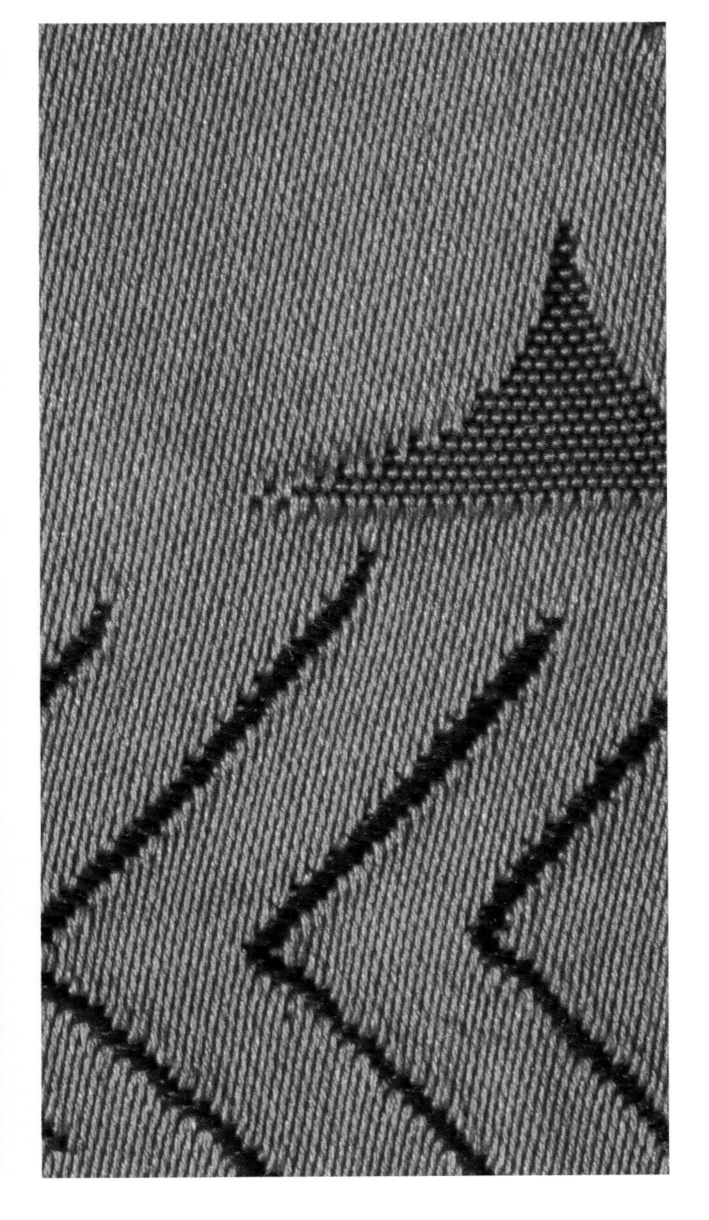

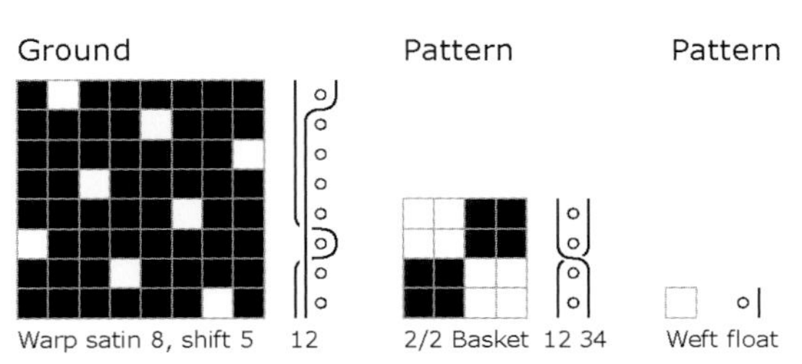

Figure 6.15. *Silk damask with self-patterning effect;* Paving Stones *by Hans Thomsson, LFS; weaves.*

In the damask *Paving Stones,* warp-faced, glossy satin 8 contrasts subtly with the grainy surface of a 2/2 basket. Note that both these effects are warp dominant, the warp satin by structure, the basket because of the warp's high sett. A third structure—a weft float—is used to draw the herringbone-like lines that "cut" the warp-faced satin, creating a strongly etched effect. The repeat of the first effect is eight ends by eight picks, the repeat of the second effect is four by four, the third effect requires one end and one pick to complete one repeat; each weave effect generates different amounts of take-up that could, after many meters, produce a difference in warp tension.

Weave effects 1 and 3 in *Swirls* are simple weaves woven with a dark blue warp and one weft series in two colors: lavender and copper. Effect 1 is an irregular rib, effect 3 a regular warp satin; the first creates a blend of warp and weft colors, while the second presents a clean-edged, smooth warp surface with a uniform color. Both are woven with all the picks of the weft. Weave effects 2 and 4 are both compound structures in which the ground weft is divided into two series with distinct functions; in both effects alternate picks weave the patterning effect on the face, while the other half of the weft interlaces in a tighter weave below to maintain the overall stability and warp-take-up. In effect 2 the lavender-colored ground picks (weft 1 in the reading note) weave a foundation in 3/1 Z twill below the copper-colored picks (weft 2) which weave as a weft satin 8 on the face of the textile. In weave effect 4, the lavender-colored picks (weft 1) appear on the surface, bound in an extended weft satin 8, while the copper-colored picks maintain a firm weave below—a modified 3/1 Z twill.

Effects 1 and 3 are uniform, warp dominant with compact surfaces, while effects 2 and 4 create the figuring in two distinct colors and by use of longer floats typical of weft patterning. No additional weft series, material, or time were used to weave this textile, which remains the equivalent of a damask or other simple textile in terms of weight and weaving time.

Like *Swirls*, *Banksia* is a damask with self-patterning wefts, woven with one warp and weft series of picks in alternate colors, black and orange. Weave effects 1 and 2 are both simple weaves. Effects 3 and 4 are compound weaves in which the odd (black) picks float on the surface, bound at intervals by a weft satin 8 in effect 3, and weft satin 16 in effect 4, while the even picks maintain a structural base below in 3/1 S twill. As with *Swirls,* the ground wefts are used as pattern wefts to add distinct, weft-faced patterning to the textile with no additional material or loom time.

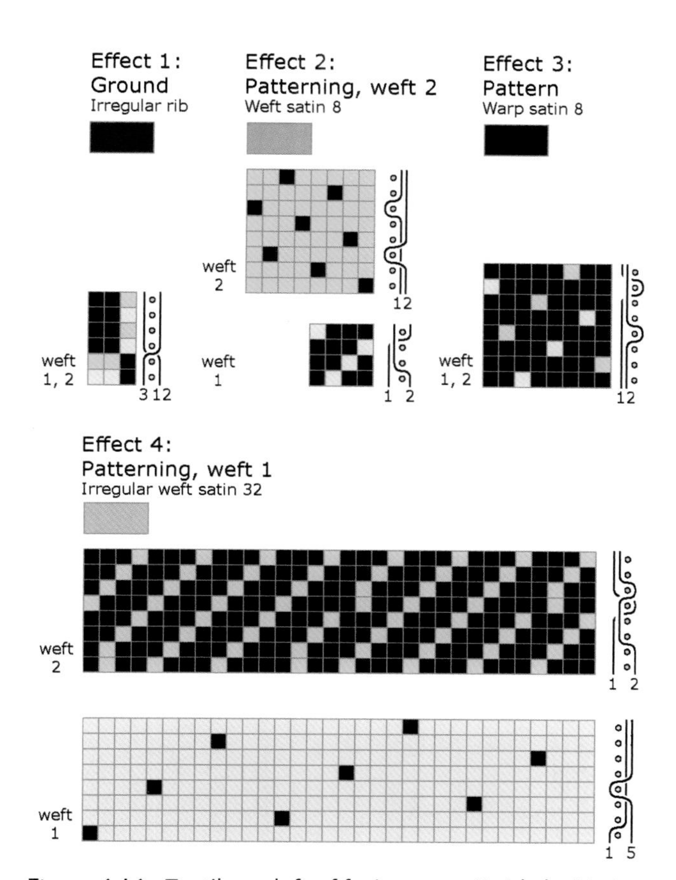

Figure 6.16. Textile on left of facing page: Swirls *by Narin Panitchpakdi, LFS, silk damask with two self-patterning wefts.* Above*: weaves.*

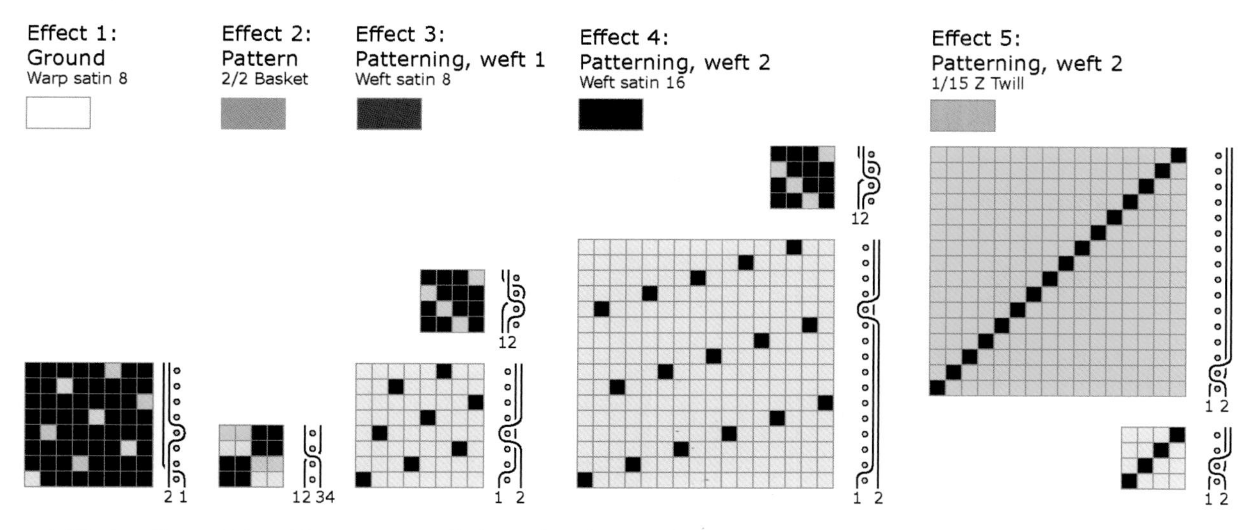

Figure 6.17. Textile on right of facing page: *silk damask with two self-patterning wefts,* Banksia *by Jennifer Robertson, LFS.* Directly above: *weaves.*

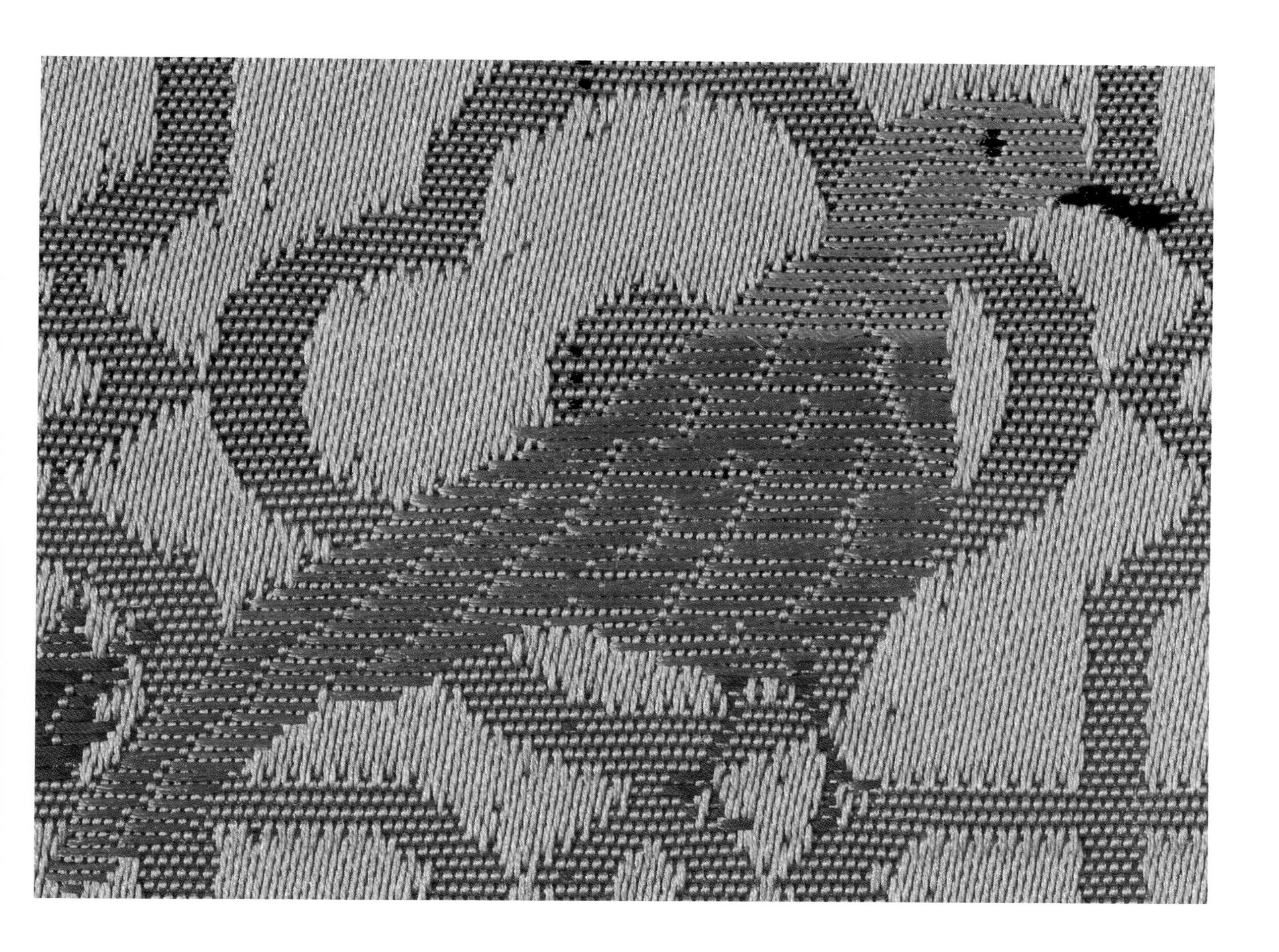

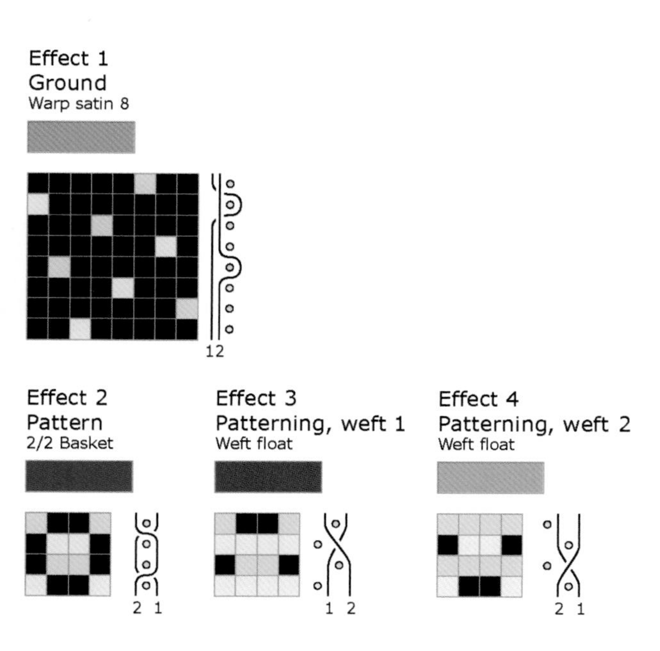

Figure 6.18. Facing page: *face and reverse of* Bhakti's Birds, *damask with two self-patterning wefts by Bhakti Ziek, LFS;* This page: *detail of face, weaves.*

Bhakti's Birds is again a damask with self-patterning wefts. Weave effects 1 and 2 are simple structures that employ one warp and a ground weft series with picks in alternate colors to create a glossy warp satin ground with wrought iron-like patterning in 2/2 basket. Both are warp dominant, the satin by structure, and the basket because of the high warp sett.

Floats of odd or even picks produce the colorful birds that stand above the damask ground. When the odd picks float on the face, the even picks remain beneath and weave with the warp in 2/1 rib to maintain a solid structure below the patterning effect. Binding points are added as needed to tack down the overlong floats and create decorative lines that define the birds' plumage. Unlike the two previous textiles of the same class, *Swirls* and *Banksia,* the colors of the alternate wefts vary over the length of the repeat, enlivening the patterning areas, while the warp's high sett and coverage maintain the uniform red of the two simple weaves effects 1 and 2.

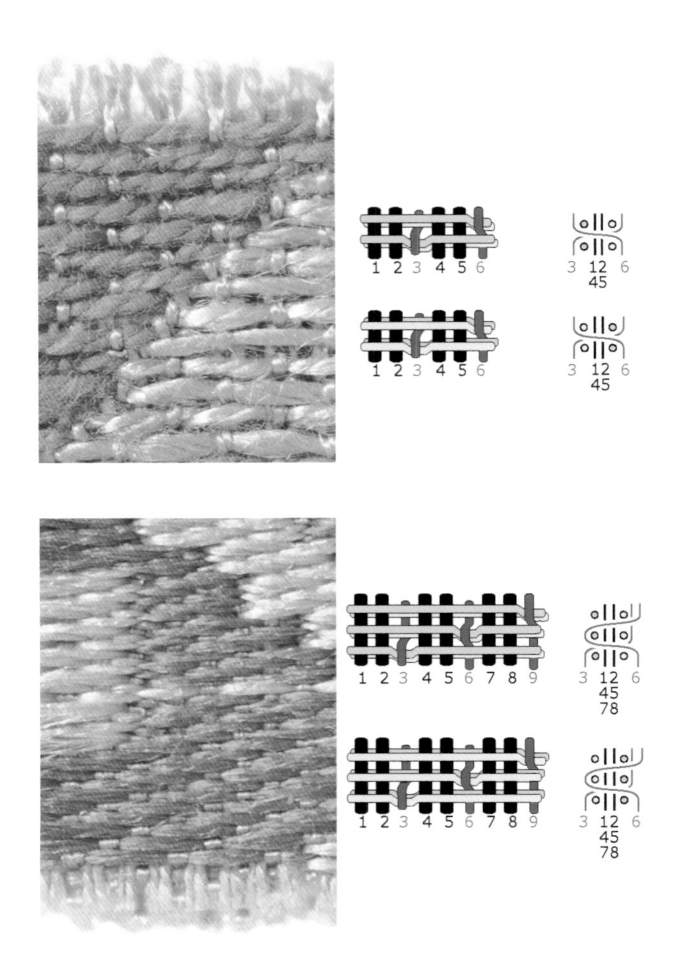

Figure 6.19. From top to bottom: *weft-faced compound tabby and weft-faced compound twill, both woven with two complementary wefts.*

Complementary Wefts

Weft-faced compound textiles employ one of the oldest techniques in figured weaving. In this class of figured textiles, a ground warp lies at the center of the cloth (at times never appearing on the face), while two or more sets of wefts are used interchangeably on the face and reverse to create both the textile's supporting structure and figuring. All weaves are compound structures, and both the face and back of the textile present patterning with equivalent binding systems, typically in plain weave or a small repeat twill. The ground warp (called an inner warp) does not interlace with the ***complementary wefts*** and merely lies between two or more weft series, which are bound with an auxiliary binding warp on both face and reverse. Figure 6.19 shows drafts and macro photographs of weft-faced compound plain weave, or *taqueté,* and weft-faced compound twill, or *samite*.

As figuring technologies evolved over the centuries, this technique was replaced by others such as lampas that made a more economical use of weaving time and materials; the ground warp became a visible component of figured textiles, patterning wefts were used to decorate the face, rather than face and reverse. Other solutions maintained the double-faced, reversible characteristic of taqueté and samite, while assigning a binding and patterning function to the ground warp.

Figure 6.20. *Detail of face and reverse,* Sheetal's Leaves.

Weft 1 on face

Weft 2 on face

1/2 Rib

Figure 6.21. *Double-faced figured weft satin* Sheetal's Leaves, *Sheetal Khanna-Ravich, LFS; hand-drawn weft sections; weaves effects.*

Today traditional weft-faced compound techniques are of interest to those who are less concerned with speed of production because of the ease with which weft materials and colors can be varied, the design changed, and patterning produced on both faces of the textile.

Hand-drafted weft sections for effects 1 and 2 of the double-faced weft satin *Sheetal's Leaves* illustrate the two main effects, which weave as weft-faced satin on face and reverse. In effect 3, the ground warp binds both wefts consecutively in plain weave to produce a 1/2 rib, which presents a warp-dominant surface on the face and reverse. Of the several samples woven for this project, various weave solutions were tested and produce slight differences between binding on the face and reverse. The yellow binding warp seen in the macro photographs in Figure 6.20 does not appear in the drafts, but as in the older technique of taqueté, binds both wefts on face and reverse in tabby.

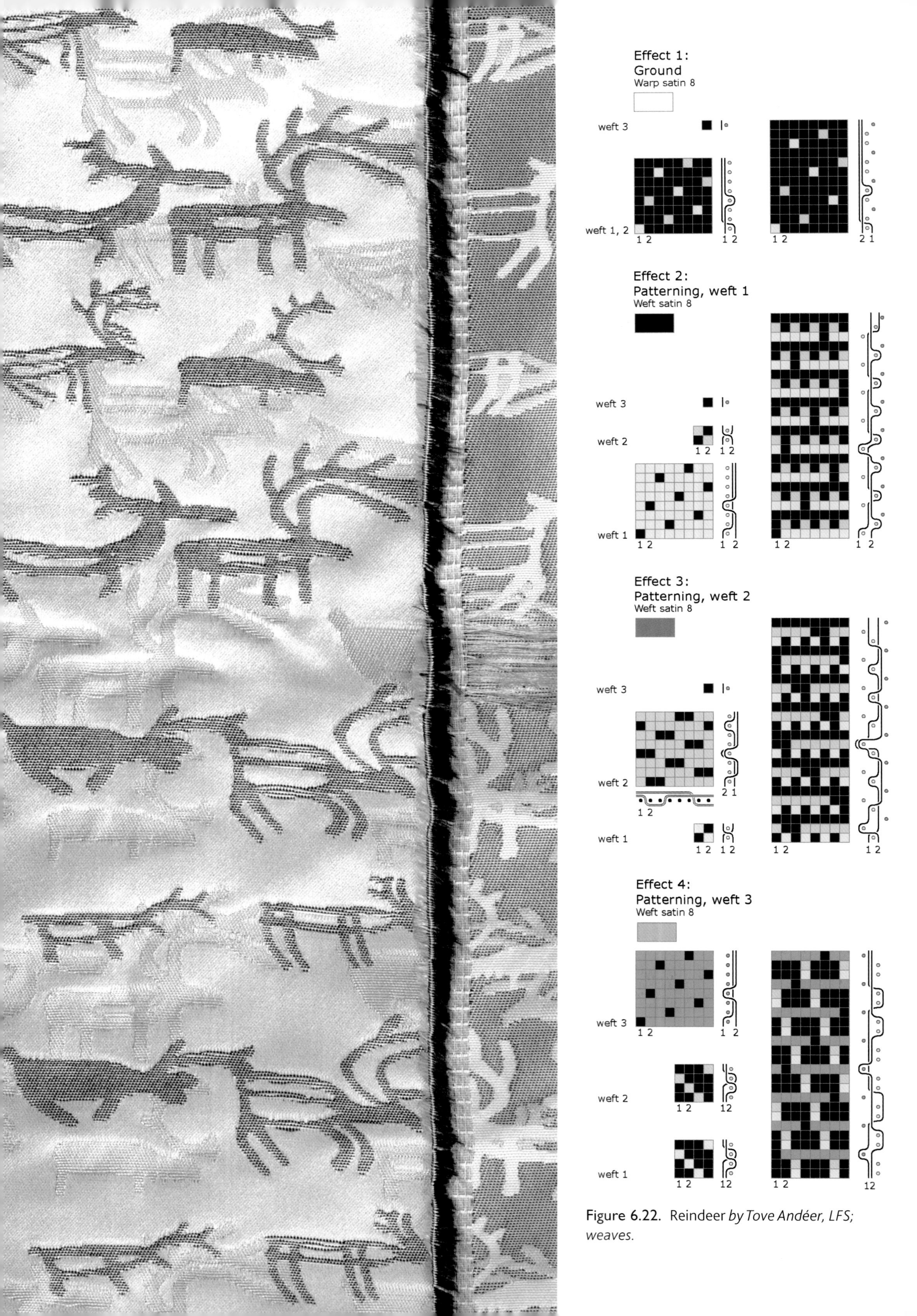

Figure 6.22. Reindeer *by Tove Andéer, LFS; weaves.*

Pattern Wefts

Pattern wefts are used to enrich the surface of figured textiles with extra elements that do not contribute to the construction of the ground weave. As pattern wefts add extra weight, weaving time, and cost to a textile, the "economical" weaver seeks other solutions first, but in the end, certain materials and designs are best woven with these extra wefts. In this book the term pattern weft refers to an extra weft series that traverses the entire warp from selvage to selvage, that is, continuous across the width. Pattern wefts may be discontinuous in the height.

Reindeer is an industrially woven figured satin with two self-patterning wefts and a ***discontinuous pattern weft.*** The ground wefts weaves alternate picks of black then white, to construct the ground, effect 1, in warp satin 8. In effect 2, the black, odd picks of the ground interlace as a weft satin above the white weft, which weaves in tabby. In effect 3, the white picks come to the fore to weave on the face in satin 2/6, while the black picks weave tabby beneath. Effect 4, a compound weave, binds the pattern weft in weft satin 8, while both ground picks weave in the same shed of a 3/1 twill. When the pattern weft is not bound on the face, it floats unbound on the reverse. The overall tightness of the compound weave holds the pattern weft securely in place. Even if the floats on the reverse were to snag or be clipped, there is no risk of the pattern weft pulling free.

Melanie's Vermicelli is a figured damask woven with a single ground weft and one ***pattern weft*** that is ***continuous*** along the length and across the width of the textile. Five of the weave effects in Figure 6.23 progress gradually from warp satin to tabby, with the pattern weft bound on the reverse. In the sixth weave effect, ground warp and weft weave as tabby, while the pattern weft floats above. Binding points are added to the pointpaper as needed to tack down any overly long floats of the pattern weft.

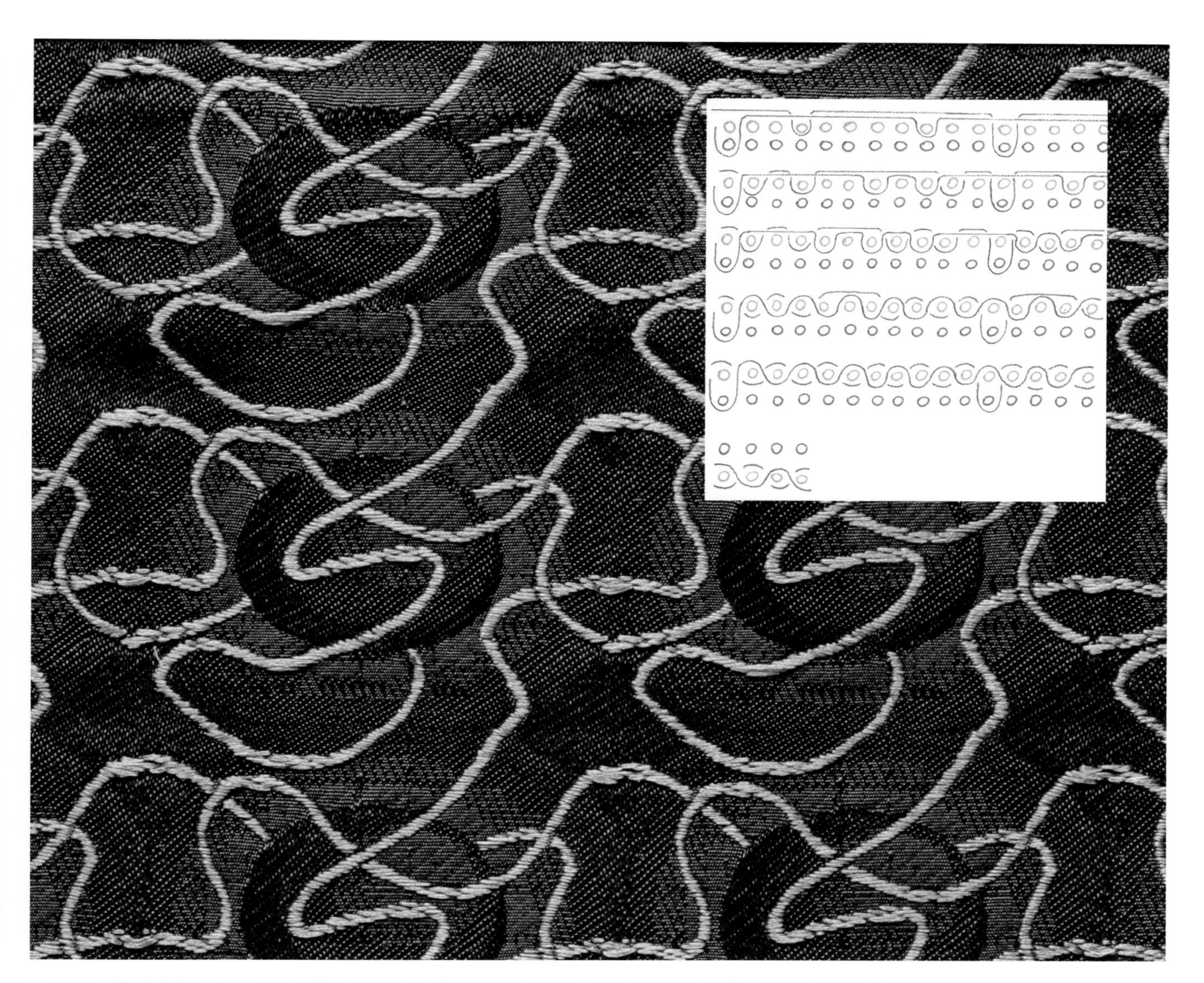

Figure 6.23. Melanie's Vermicelli, *damask with patterning weft and weave drafts by Melanie Olde, LFS.*

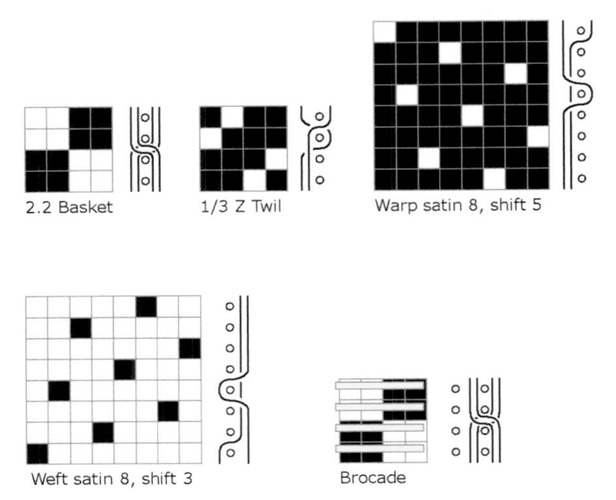

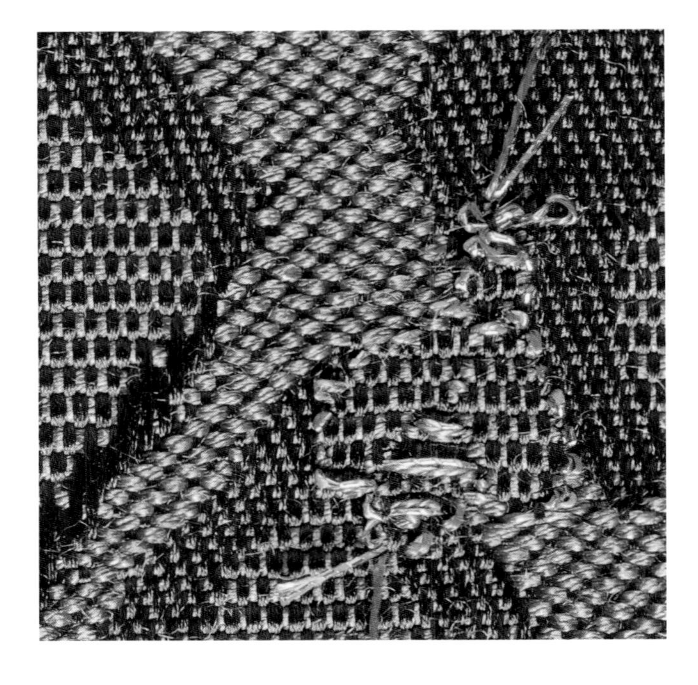

Brocading Wefts

Brocading wefts decorate a woven textile in much the way embroidery does, only instead of a needle, a small shuttle serves to pass each weft color above and below the surface of the cloth. As with embroidery, the point where a thread passes from the face to the reverse and returns to the face may present an unattractive stitch; for this reason most brocades are woven face down, to leave any unsightly binding points or unwanted floats on the reverse. A true brocading weft is discontinuous across the textile's width and does not traverse a textile from selvage to selvage, being limited to the width of a motif or cluster of motifs. The insertion of a small shuttle for each distinct weft color, passed in its own shed, reduces weaving speed. Now, as in the past, brocading tends to be reserved for figuring with precious materials, such as gold, or materials that if bound on the reverse would produce a defect on the textile's face. Brocading wefts may be employed in lieu of pattern wefts when figuring is limited to small areas and requires multiple weft colors, as brocading economizes on material and avoids adding the weight of extra wefts to the textile.

Honeycomb is a brocaded damask. Note how all effects are woven with simple structures, with the exception of the brocaded motifs. These are woven with a gold-foil-wrapped weft that floats above a 2/2 basket ground; a few binding points tack the gold weft down and define the shape of the bee's wings and head. Note that the brocading is limited to the small figures, and does not float on the reverse between motifs.

Figure 6.24. *Brocaded silk damask,* Honeycomb *by Martha Porter, LFS; weaves; detail of reverse.*

Green Triangles is a damask with two self-patterning wefts (blue and yellow) and three brocading wefts (pink, violet, and lavender). The damask figuring in warp satin 8 contrasts with the tighter, blended surface of a 2/2 basket ground effect. Yellow, alternate picks of the ground weft float unbound on the face to delineate the triangular motifs, while the blue picks bind below in 2/1 basket. Floats of the blue ground picks form the dots scattered on the textile's ground, while the yellow picks interlace with the warp to maintain the structural foundation beneath the floating wefts. Three brocading shuttles are used to weave the pink, blue violet, and lavender dots. Because of the proximity of the multicolored brocading, each weft series requires a separate shed.

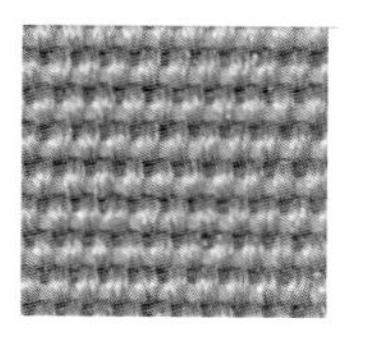

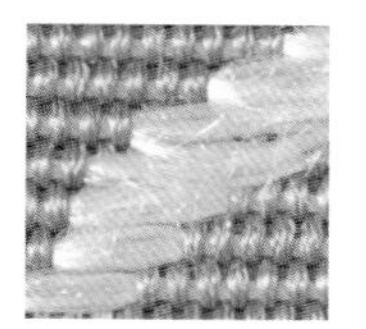
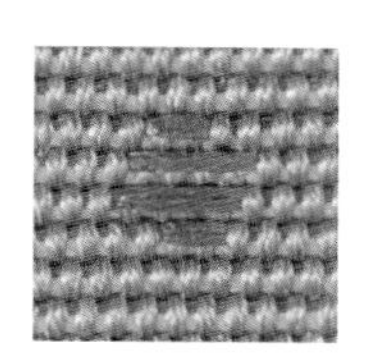

Figure 6.25. *Detail of the five effects;* Green Triangles, *brocaded silk damask by Elisabeth Egger, LFS;* Below: *reverse and face.*

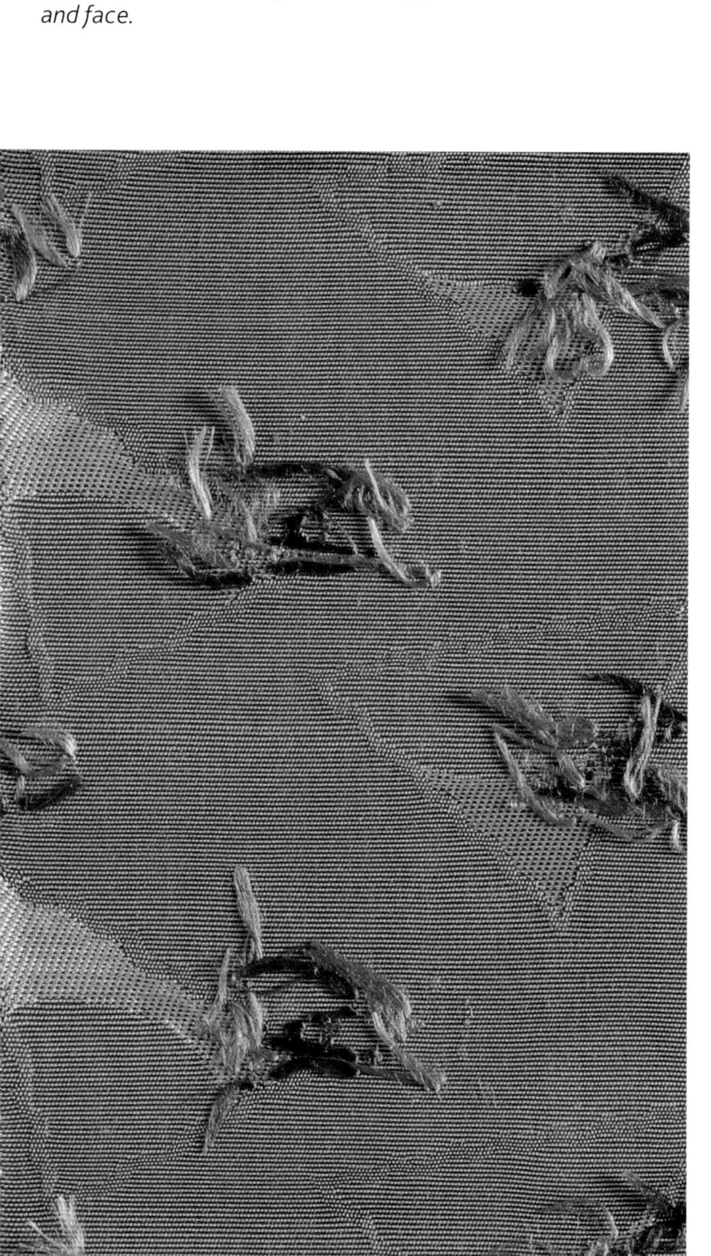
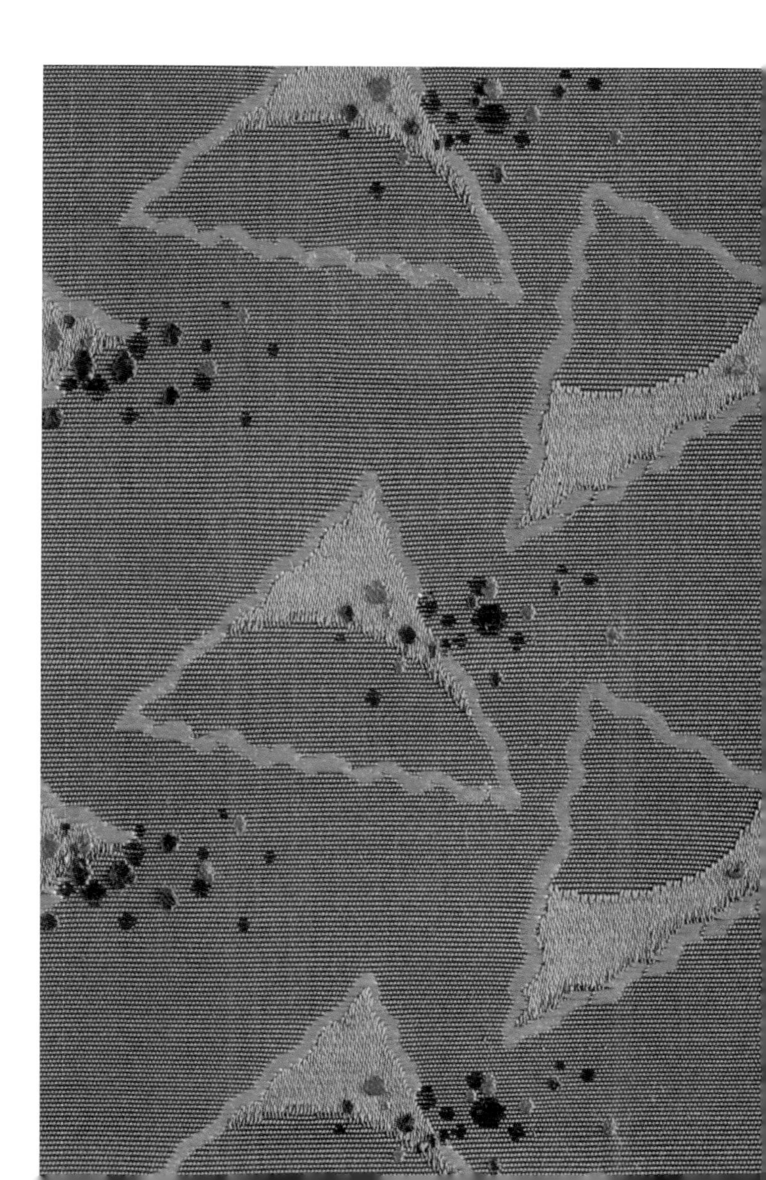

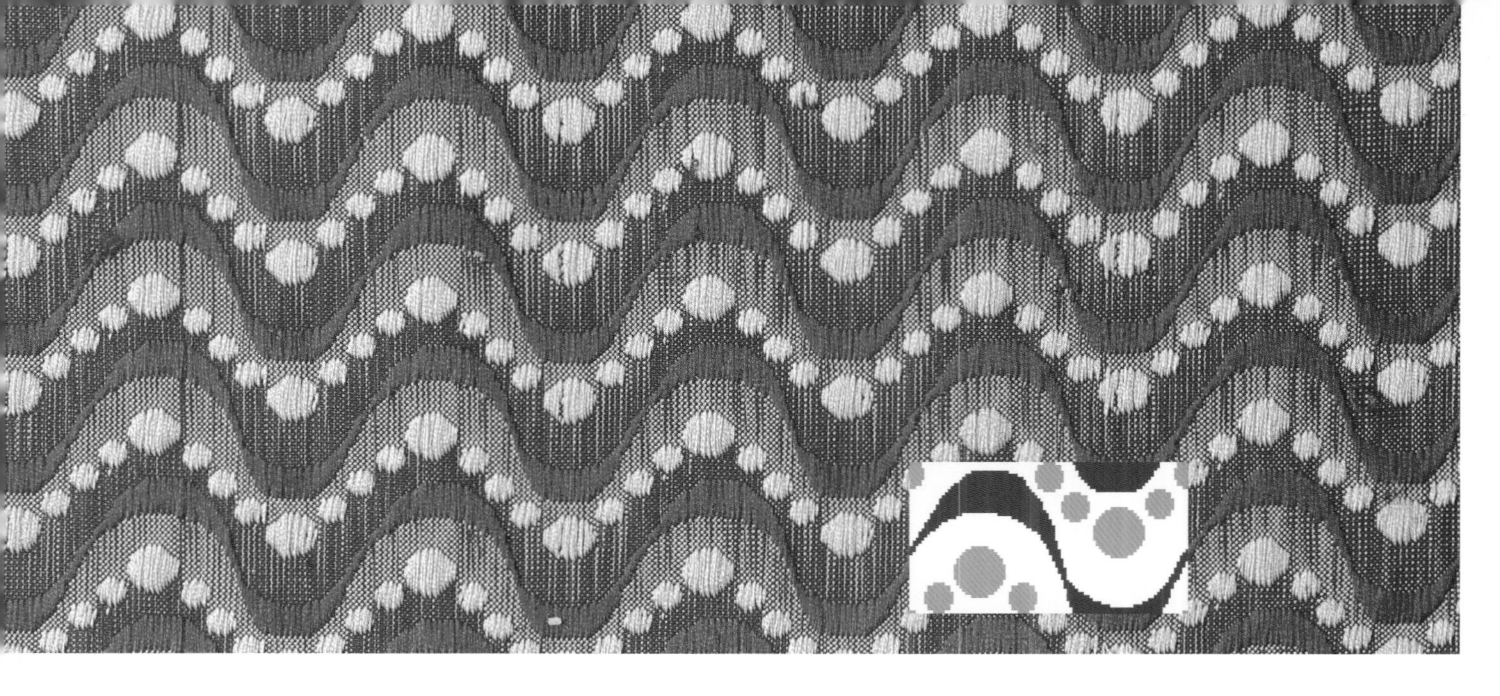

Warp patterning

Self-patterning Warps

Like self-patterning wefts, self-patterning warps are ground elements that are employed for a double function: to create the textile's supporting weave structures and patterning. As with self-patterning wefts, the use of a ground warp to perform patterning functions is economical, and does not add extra weight or warping time to figured cloth.

Haitienne is a contemporary remake of an older technique of the same name that in its original form was woven with multiple warps. The structural solution adopted by the designer obviates the need for a pattern warp series. In the ground weave, effect 1, alternating white and red warp ends weave together as a 2/1 basket (called *louisine*); in effect 2, all the odd-numbered, white ends float on the surface to create the figuring, while the red, even ends weave as tabby beneath. In effect 3, the even-numbered, red warp ends are raised to produce the figuring, while the white ends weave as plain weave to maintain structural strength below. One warp series in two colors and a single weft series create three distinct weave effects.

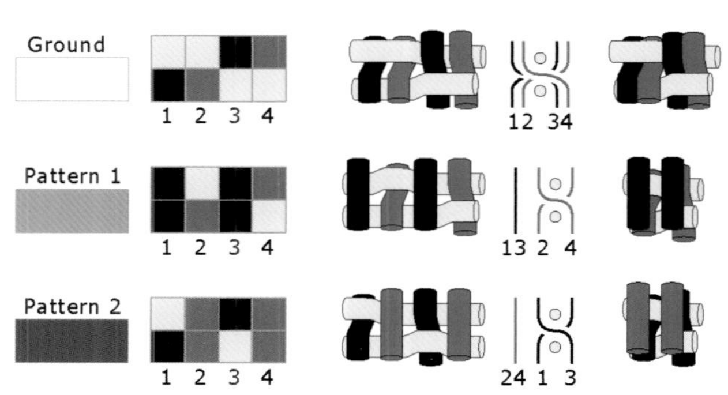

Figure 6.26. Top: *Figured lousine with two self-patterning warps, pointpaper design,* Haitienne *by Martin Ciszuk,* LFS; Middle: *weaves;* Bottom: *detail of face and reverse.*

Pattern Warps

Pattern warps are extra elements used to add figuring to a ground textile. Because of their decorative nature, figuring ends are more loosely bound than ground ends, and therefore have a lower take-up factor. For this reason pattern warps are usually wound on a separate beam. Their use enriches the surface, but also adds time to the warping process and extra weight to a fabric.

If the pattern ends are widely spaced and few in number, they may be mounted on a creel, as in the case of *Fingerprint,* a figured 1/2 rib with self-patterning warp and weft, and ***discontinuous pattern warp*** and ***weft.*** The components of each effect—ground and pattern warps, two ground wefts, and pattern weft—are drafted separately as drawdowns with relative weft sections. Note how in all effects, except the third, the ground warp interlaces as plain weave or a derivative. The tightness of these structures gives relief to the figuring, whether woven with extra patterning elements, or self-patterning ground ends and picks.

Figure 6.27. *Detail, weave drafts, and 1:1 photograph of* Fingerprint *by Anne-Birgitte Hansen, LFS.*

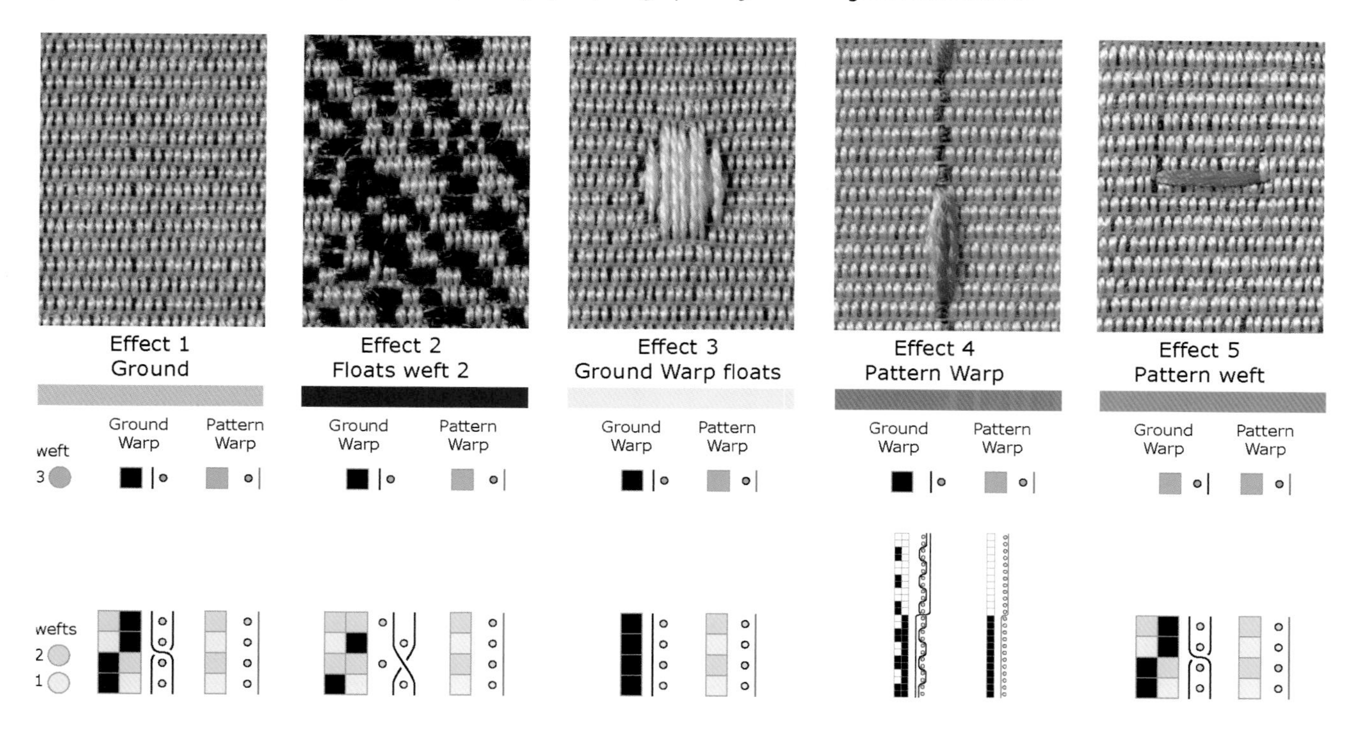

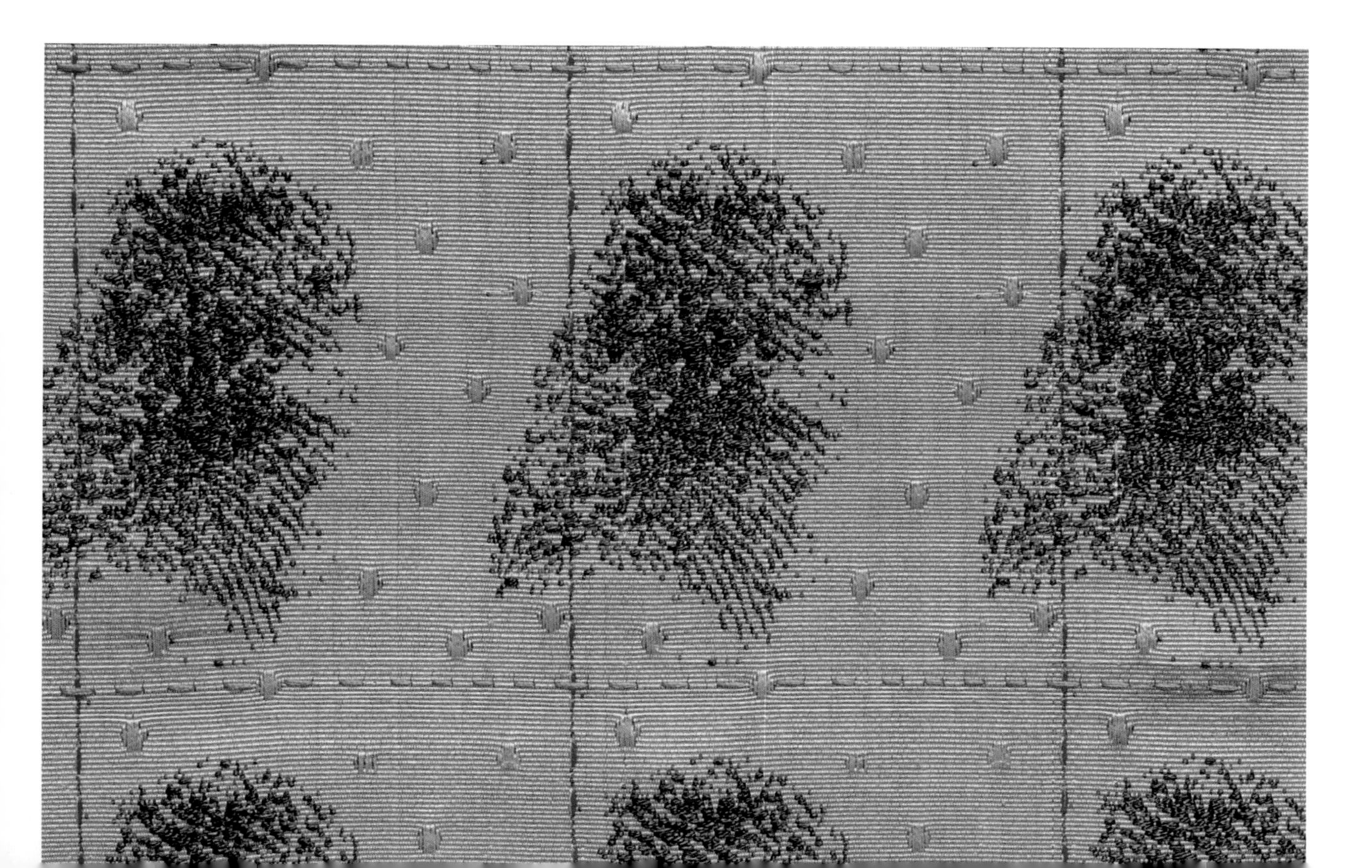

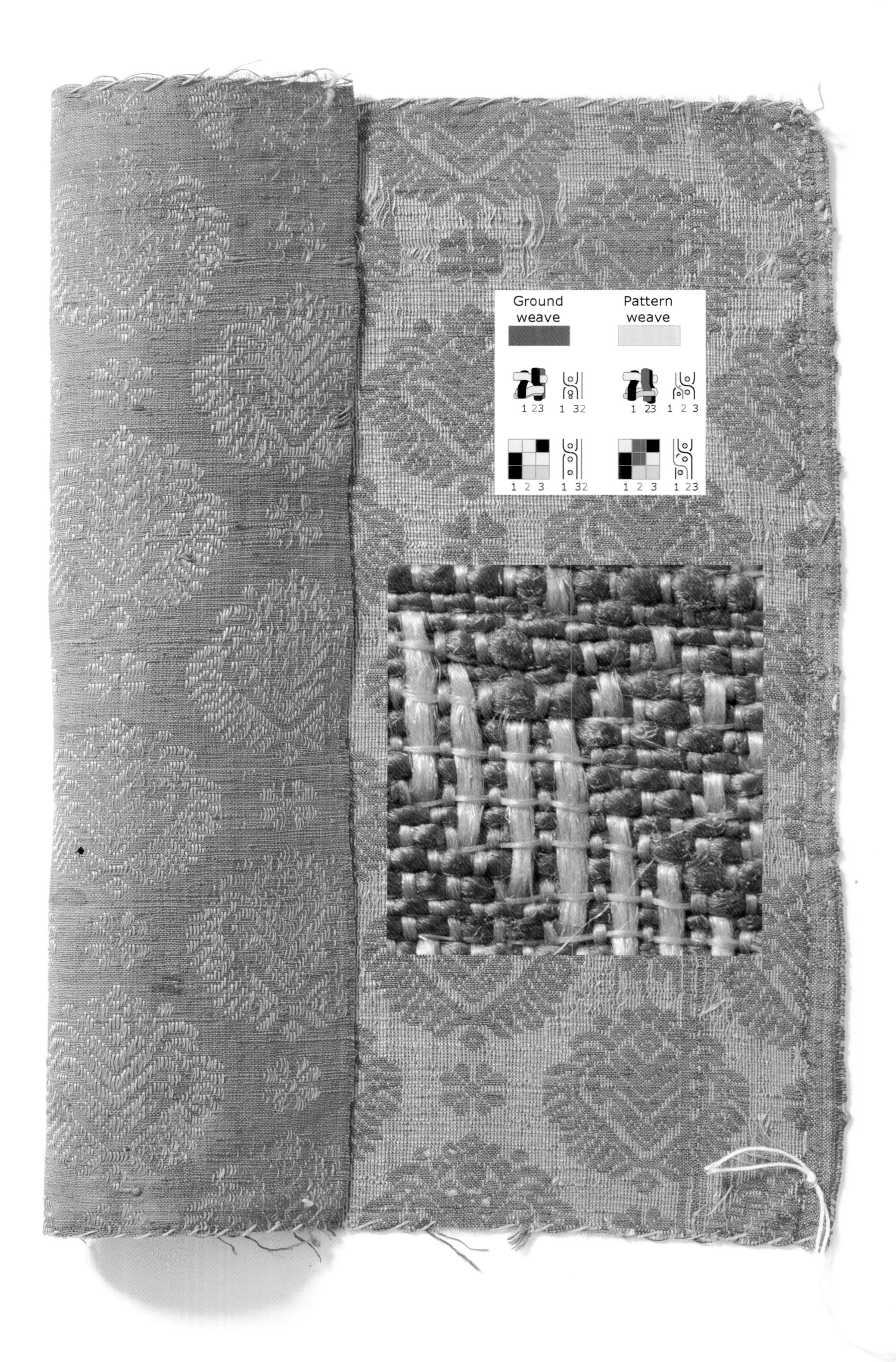

Ground weave
Pattern weave
1 23
1 32
1 23
1 2 3
1 2 3
1 32
1 2 3
1 23

The warp-patterned silk in Figures 6.28 is woven with one ground and one continuous pattern warp series, one ground and one binding weft. The proportion between ground and pattern warp ends is 2 to 1; the proportion between ground and binding wefts is also 2 to 1. Drafts and macro photograph illustrate the relative dimensions of ground and patterning warps, and ground and binding wefts; these produce a weft-dominant ground and a patterning effect in which loosely twisted figuring ends cover the underlying tabby ground. Whether brought to the face, or left on the reverse, take-up of the patterning ends is uniform.

Figure 6.28. Facing page: *plain weave ground patterned with a pattern warp, bound in tabby by a fine binding weft. Courtesy of LFCHT.*

Bauhaus, Bolsterlang, and *Big Vermicelli* belie the notion that multiple patterning elements add to the complexity of a figured textile. In all three textiles a ground warp and weft weave as tabby to create effect 1, while three pattern warp series float on the reverse; in turn, pattern warps 2, 3, and 4 float on the face to create pattern effects 2, 3, and 4.

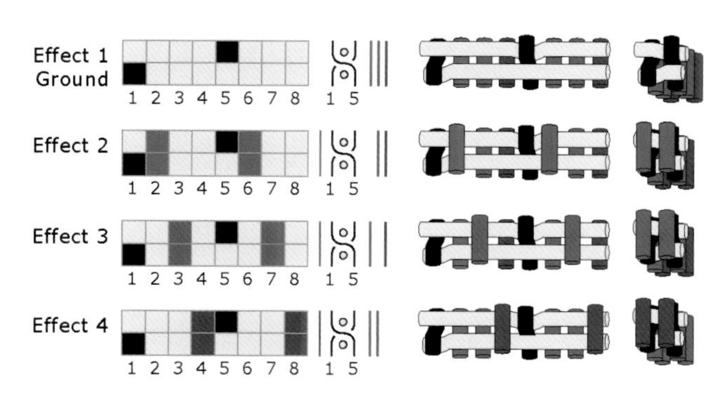

Figure 6.29. Top from upper left to lower right: Bauhaus *by Gregg Conly,* Bolsterlang *by Berthe Forchhammer,* Big Vermicello *by Ulrikka Mokad, LFS. Weaves.*

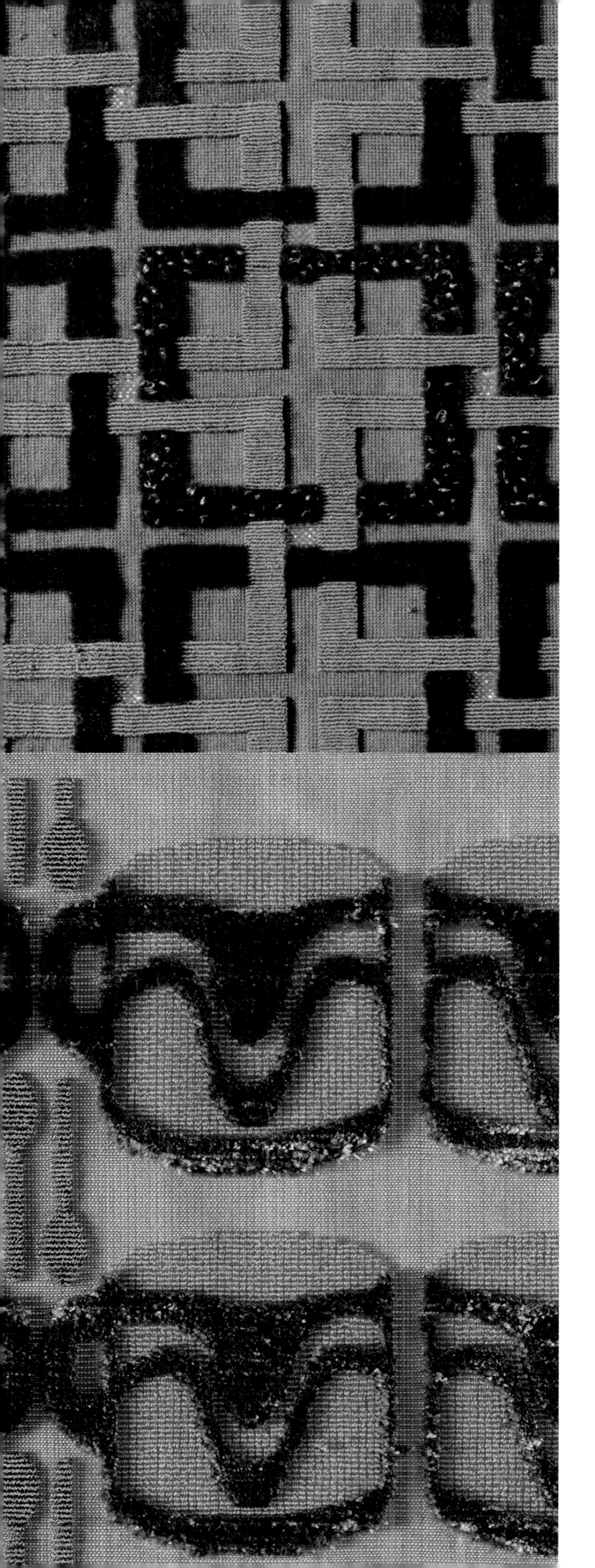

Figured Velvet

Velvet concludes this section on warp-patterned textiles. The distinguishing characteristic of velvet is the pile, a series of loops or tufts that pattern the textile's surface. In all other techniques presented in this chapter, warp and weft interlace to construct a two-dimensional plane, whereas in velvet, the ends of a special patterning warp, a *pile warp,* sit roughly perpendicular to the ground, producing a three-dimensional textile. Typically, between ten to fifteen tufts per centimeter (twenty-five to thirty-eight per inch), across width and length, decorate the surface of the velvets in this book. Given the relative coarseness of velvet (with so few patterning units per linear centimeter or inch), imperfections of design or execution become highly visible. Velvet is difficult to weave, demanding to design—those interested in working with this technique should be familiar with its particular construction and tools.

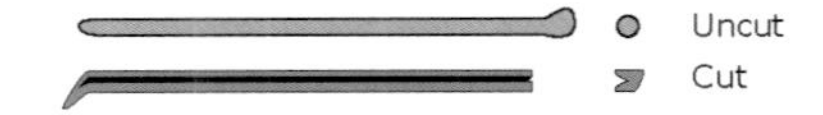

Figure 6.30. *Rods for uncut and cut velvet.*

Pile tufts, uncut or cut, are produced as the velvet is woven by placing velvet rods, pictured in Figure 6.30, beneath raised pile ends to form loops. The uppermost rod, cylindrical in section, is used to create uncut loops. Placed in the shed above the ground warp and below a selection of figuring ends, it remains in place for several successive ground picks, until each loop is firmly cinched into the ground weave; only then is the rod pulled free and placed in a new shed to create another row of loops. The lower rod in the same figure is used to create cut tufts. The rod is placed groove upward into the shed; loops are formed and tightly cinched into the ground, the point of a razor blade runs along the groove, cutting free the rod, which is placed in a new shed to create the next row of tufts.

The construction of the ground weave, whether plain, twill, or satin, must ensure that the loops or tufts that form the patterning do not pull free from the finished cloth. In the drawdowns and sections shown on the following page, note how pile ends, cut or uncut, are bound in a vise between ends and picks of the ground weave.

Figure 6.31. Top: Piazza San Marco—Day, *giselé velvet with gold brocading and alluciolato, Jan Paul.* Bottom: Cups and Spoons, *two-pile ciselé velvet, Hannah Yate, LFS.*

Figure 6.32. Facing page, from top to bottom: Gingko, *uncut velvet, Elizabeth Tritthart; two-pile uncut velvet, Narin Panitchpakdi; two-pile uncut velvet, Tove Andéer, LFS.*

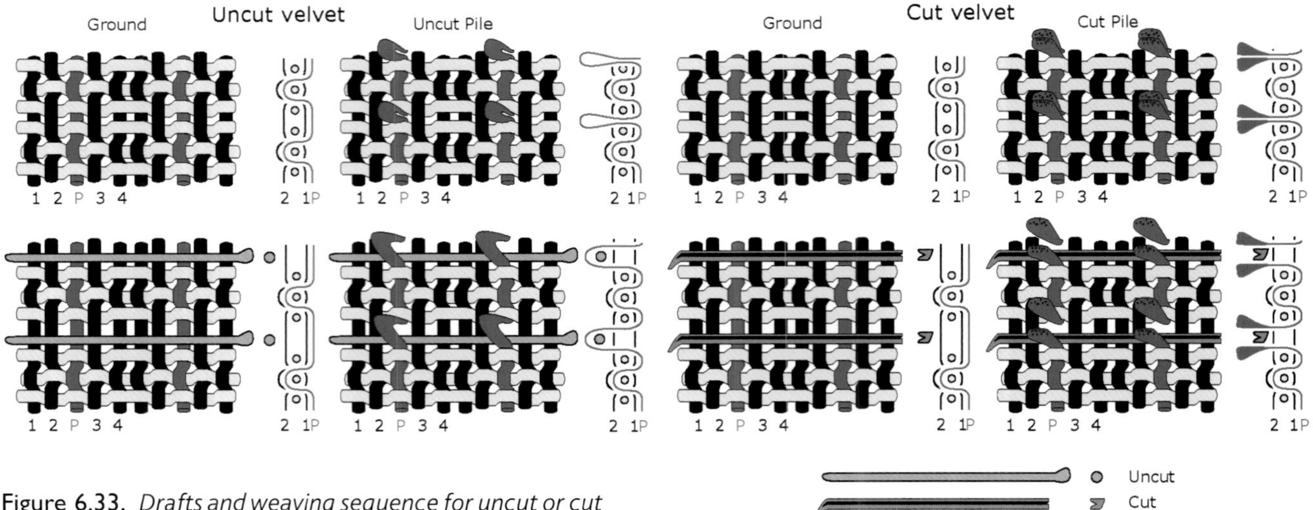

Figure 6.33. *Drafts and weaving sequence for uncut or cut voided velvet.* Right: *Rods for uncut and cut velvet.*

The simplest techniques to weave are uncut velvet and then cut velvet. Figure 6.33 illustrates how ends pass over the rod to create loops, or remain below to create the ground, or voided effect. Depending on the rod employed, uncut or cut velvet is produced. Design possibilities are expanded if a second series of pile warps is added to the first. When weaving, one or the other pile warp series is raised over a rod to create the patterning. Two-pile velvet may be woven as uncut or cut.

Giselé velvet, or cut and uncut velvet, is woven with two sets of rods inserted in a precise order. A cylindrical rod is inserted in a first shed; the shed is closed, and then a second shed is opened and a grooved rod inserted. After several ground picks are woven, the grooved rod is cut free to produce tufts and the cylindrical rod is pulled out to create uncut loops.

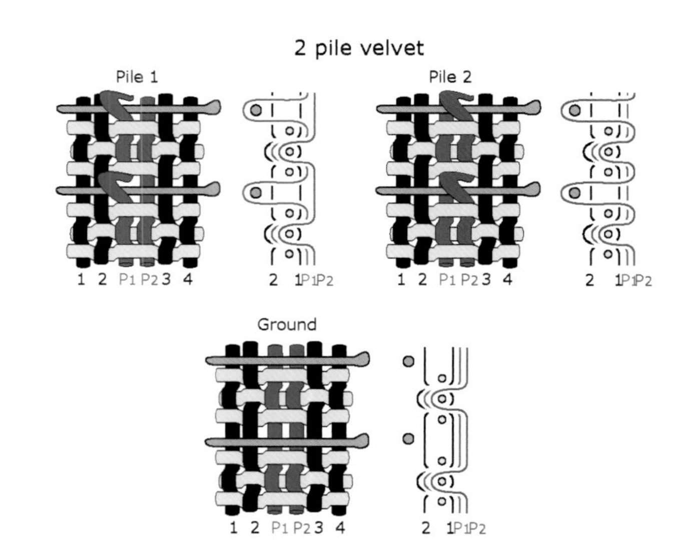

Figure 6.34. *Drafts for two-pile uncut velvet.*

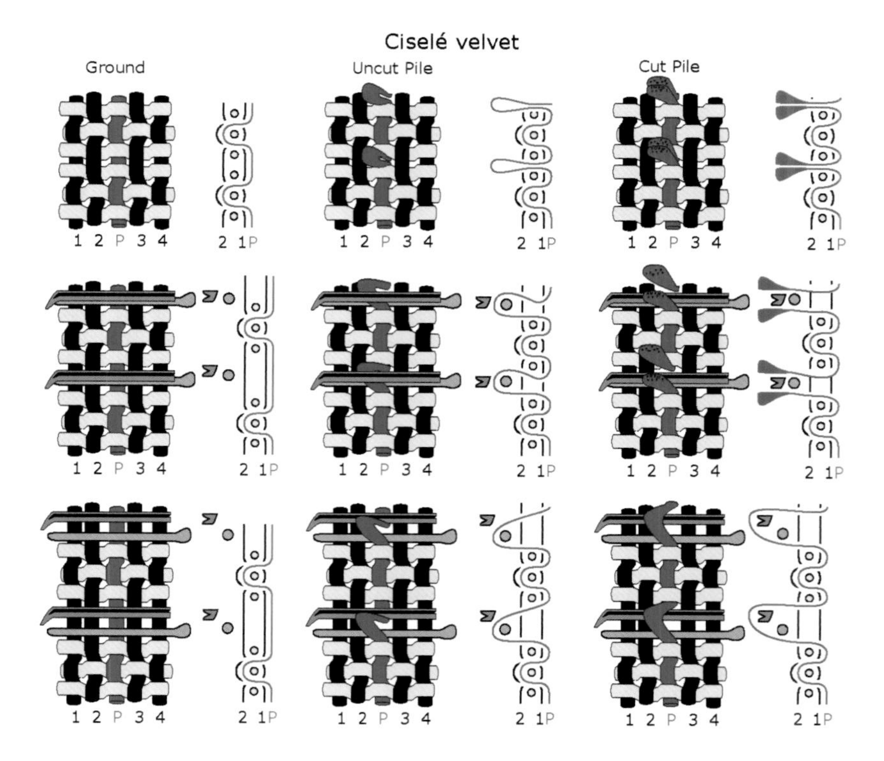

Figure 6.35. Left: *Drafts and weaving sequence for giselé velvet.*

Figure 6.36. Facing page, left, from top to bottom: Vermicello, *uncut velvet, Ulrikka Mokad; two-pile uncut velvet, Päivi Fernström; two-pile giselé velvet with pattern weft, Hans Thomsson.* Right: *detail* Vermicello, *LFS.*

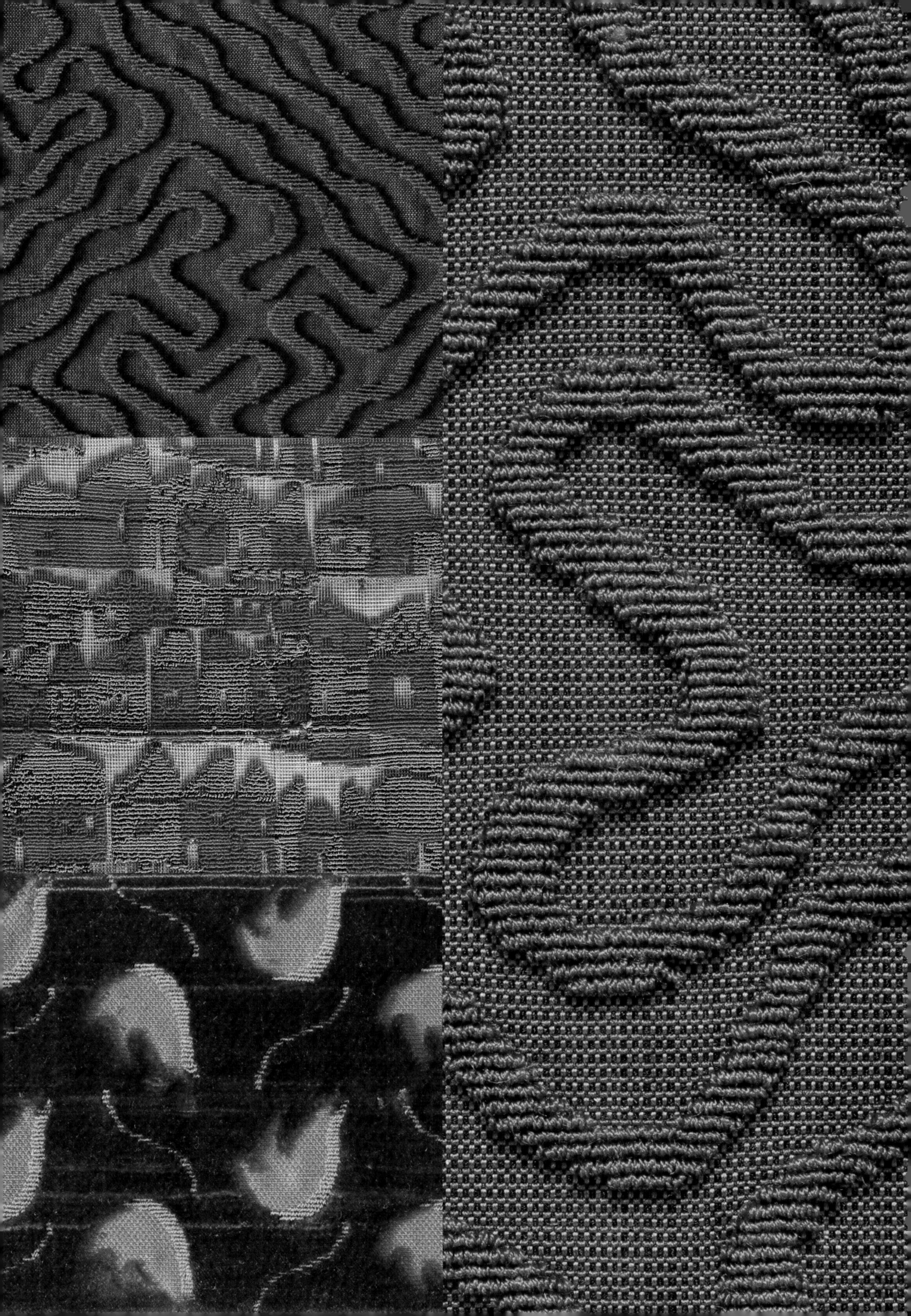

Multiple cloths

Multilayered fabrics are easily woven on shaft looms, but experimentation with this large and varied class of textiles is greatly facilitated by weave-patterning technologies. Examples in this book range from plain weave doublecloths to lampas and figured velvets woven as multiple layers. The ratio between ends and picks of each layer may be 1:1 or unequal; layers may be stitched together where these exchange, or "sewn" together with a weave or extra binding points.

Alternate Doublecloth is handwoven on a fil à fil warp in black and white; weft colors vary along the textile's height where weave effects exchange. Of the six effects employed to weave this doublecloth, the first two are simple twills in which all ends and picks weave as a single layer; the remaining weaves, twills, and tabbies weave as detached layers. In all the double-layered effects, the ratio of ends and picks of the upper and lower layers is 1:1.

Differences in interlacement density between effects produce interesting yarn deflection and lend depth to the surface of the cloth. Note the distribution in the pointpaper design of weave effect number six and observe the relative drawdown and weft section. The lower layer interlaces as tabby, while the face layer weaves as 1/3 twill. The take-up difference between the two layers, combined with the distribution of the white in vertical columns, produces yarn deflection that adds depth to the surface of the textile, but would generate unequal warp shortening if a longer piece of the textile is woven.

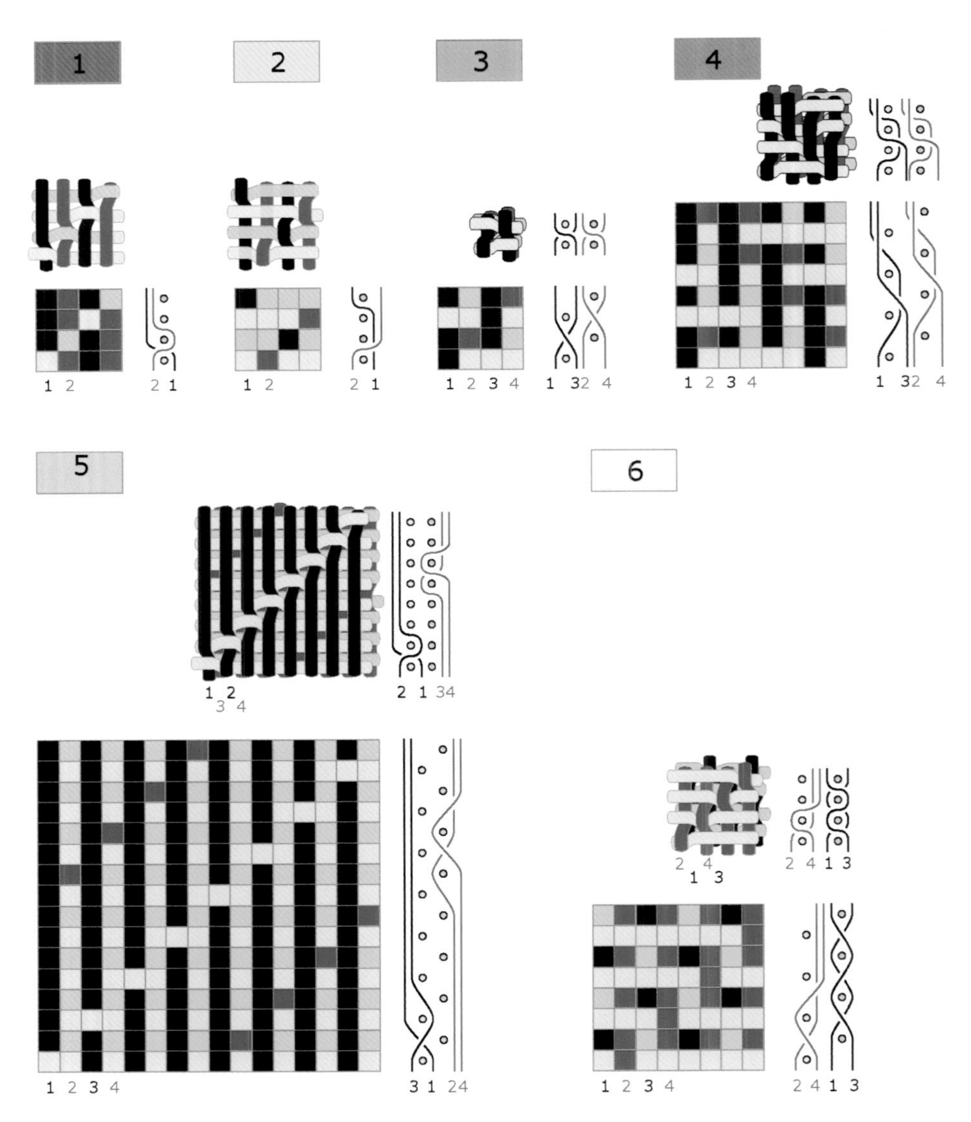

Figure 6.37. *Drafts, detail, pointpaper design, and photograph of* Alternate Doublecloth *by Susanne Heindl, LFS.*

Lampas

Lampas encompasses a wide variety of techniques, some simple, others complex. All forms of lampas have in common an auxiliary warp that serves to bind pattern wefts when these appear on the face of the textile, and again on the reverse to limit float length. The handwoven lampas seen in Figures 6.38 is an example of a simple lampas with a proportion between ground and binding warps of three to one. Ground warp and weft interlace throughout the cloth as a 5/1 S twill, while a binding warp weaves as a 1/2 S twill with a gilt paper pattern weft. Ground and patterning effects are the result of the position, above or below, of the ground structure relative to the pattern weft. As the pattern weft traverses the loom's width, the ground warp is lifted in mass to create the ground effect and lowered in mass where the patterning effect appears on the surface.

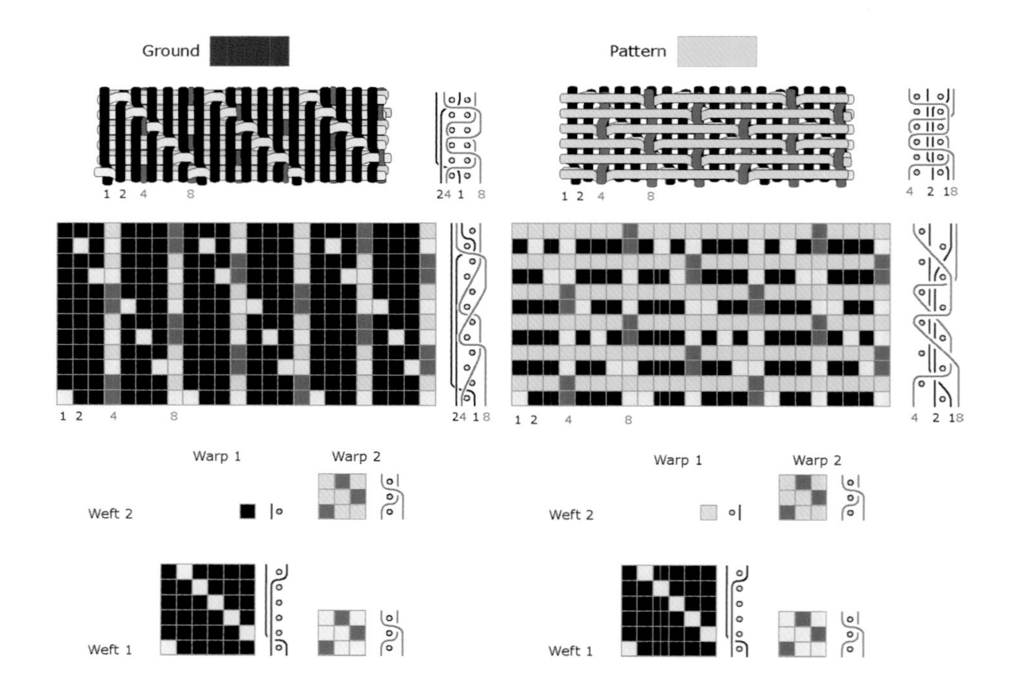

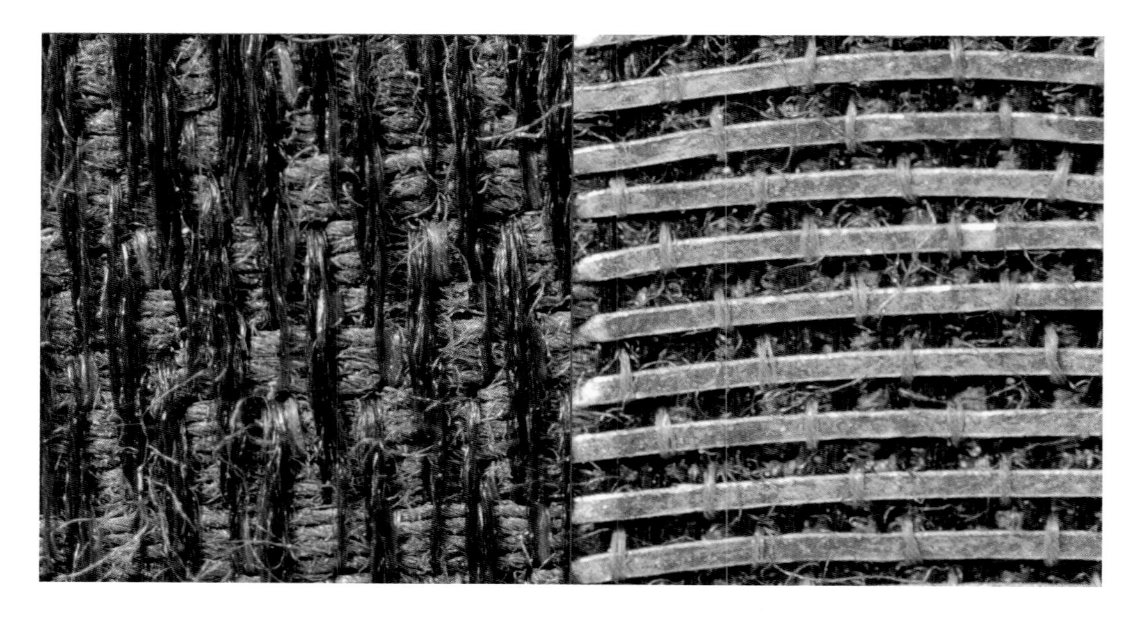

Figure 6.38. *Weaves, details of ground and pattern effects, and photograph of a Japanese lampas with gilt paper pattern weft, CJH.*

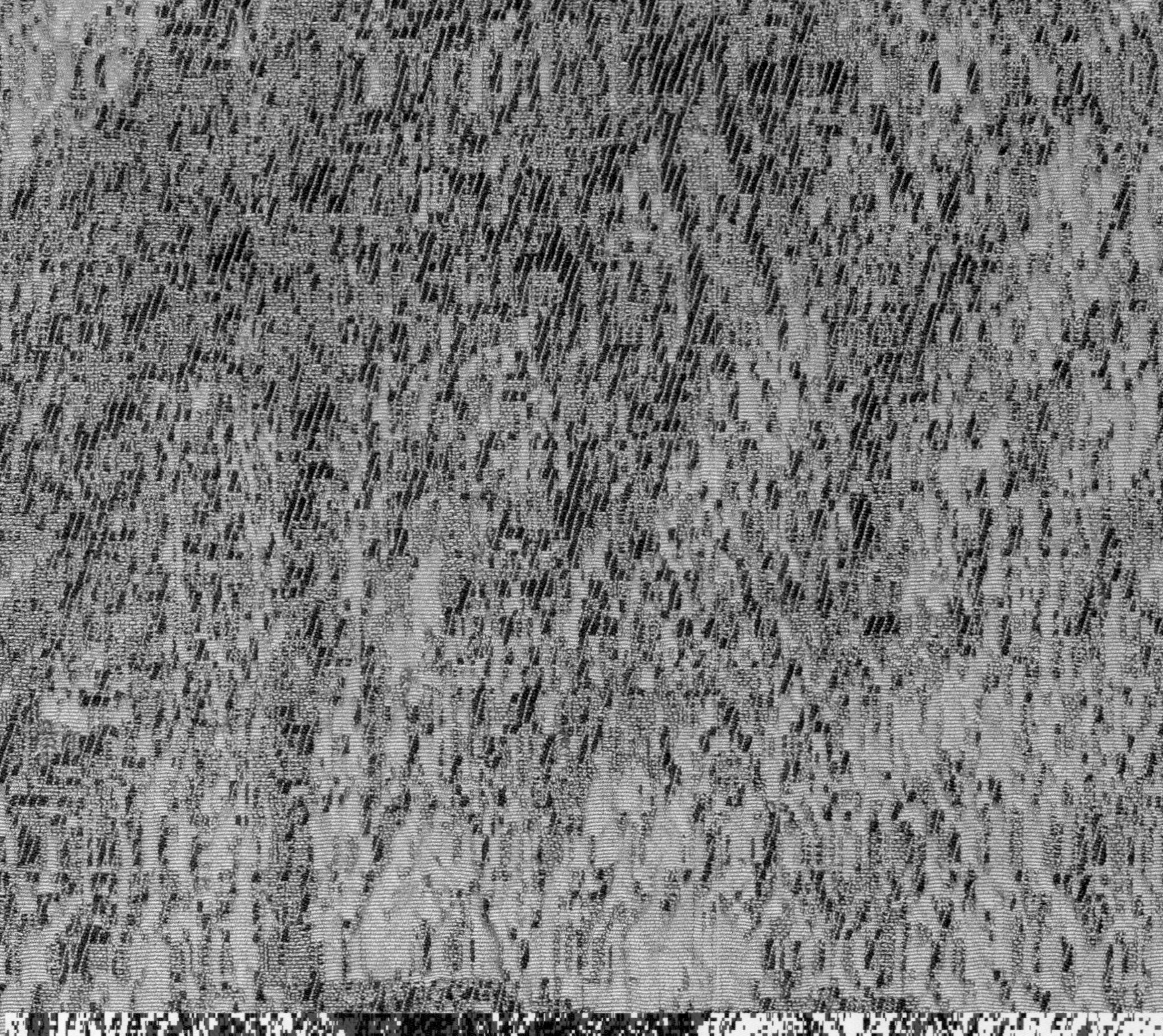

This second, industrially woven lampas is more complex than the previous example. Effect 1 presents a tabby ground; the other three effects are weft-faced structures, each with a distinct surface. In effect 2, the first weft series floats unbound; in effect 3, the second weft series is bound by the ground warp in satin 8; in effect 4, a third weft series is bound by the ground warp in 1/7 Z twill. Note that the auxiliary warp never binds the wefts of the upper layer and is limited to binding the wefts that compose the lower layer of each effect. Effects 1, 3, and 4 weave as doublecloth with separate layers. In the second effect, the two layers are stitched together where weft series two binds with the warps of both layers.

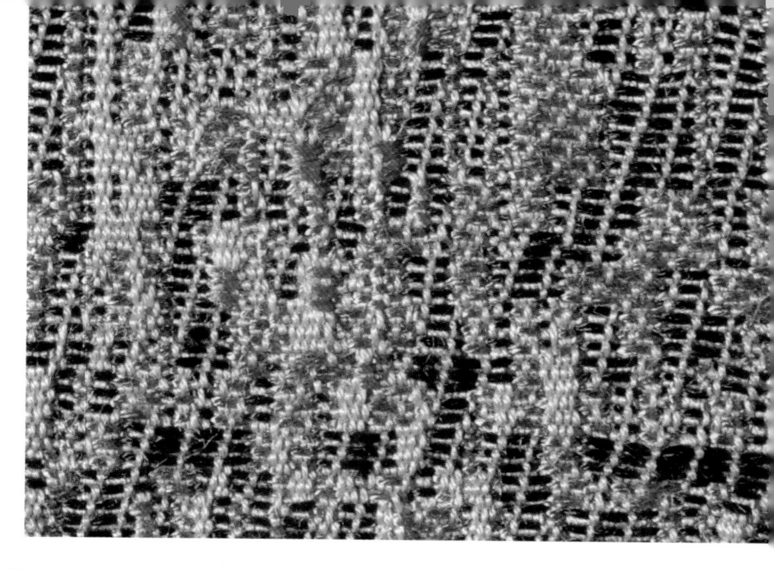

Figure 6.39. *Industrially woven doublecloth lampas with one ground weft and two pattern wefts,* Fragment *by Berthe Forchhammer, LFS. Photograph and pointpaper* (facing page), *reading note* (below), *and detail* (above).

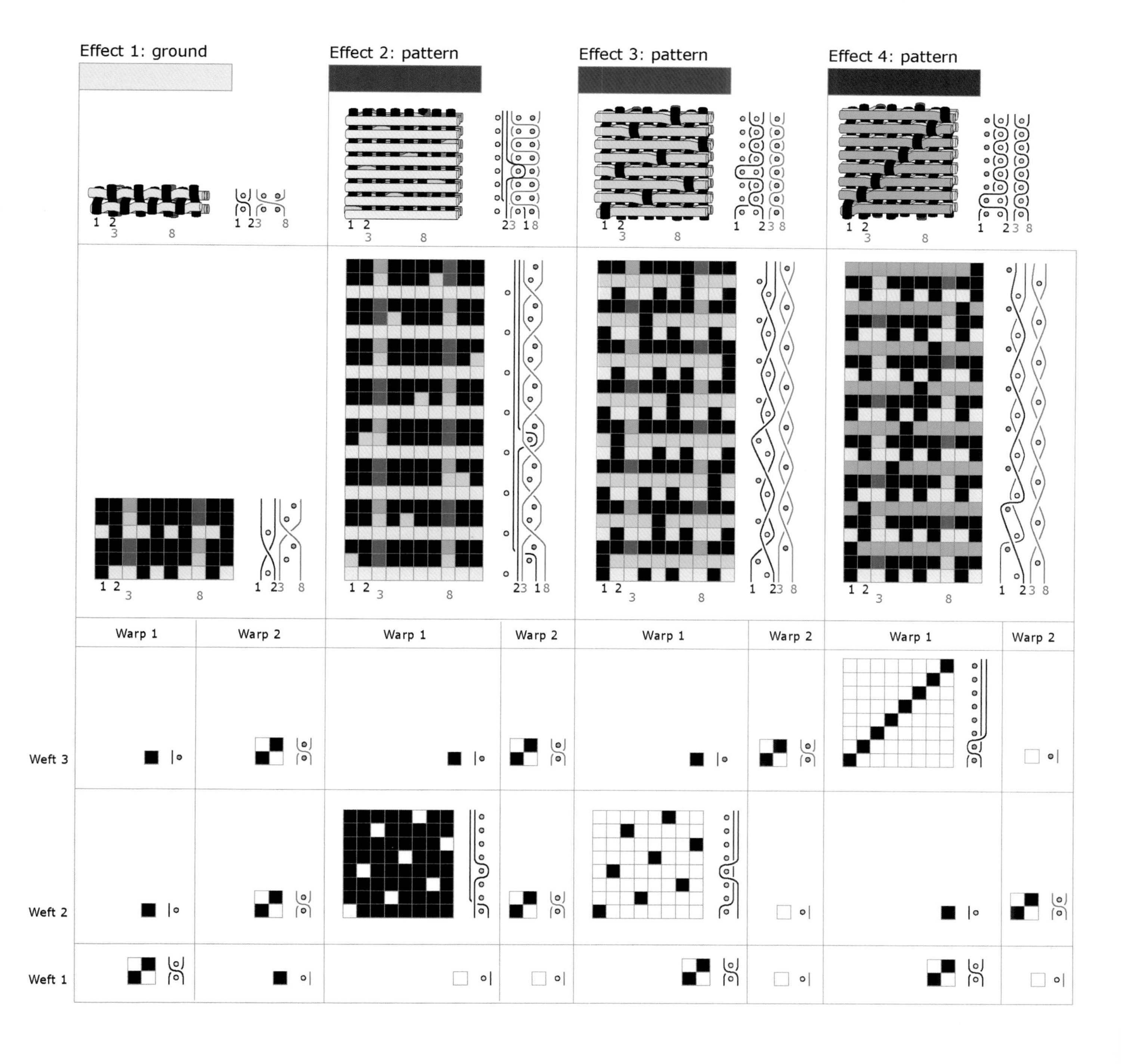

Part III

Woven Structure, Design, and the Jacquard Medium

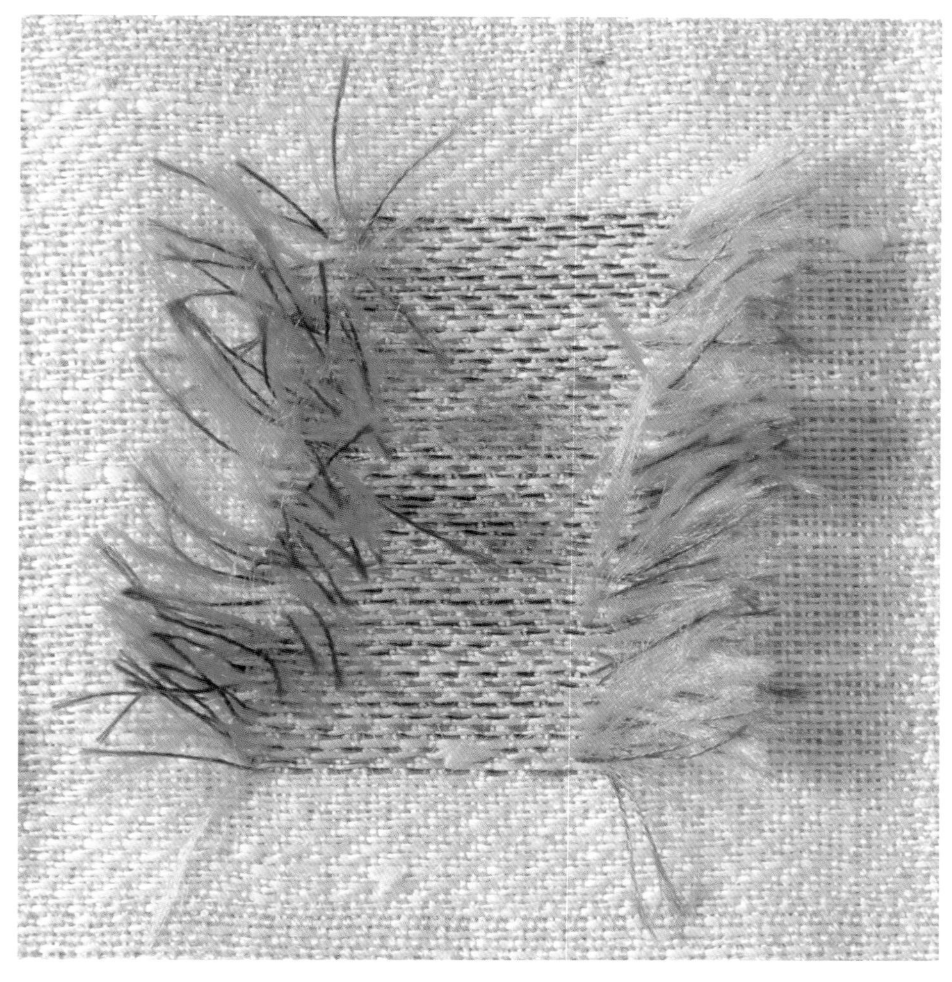

Clipped Squares *by Helena Loermans, LFS.*

7 Weave glossary, choosing and building weaves, evaluating weaves

The ability to build weaves that produce a desired aesthetic and/or functional effect is one of the basic tools of woven design. In addition to a strong understanding of weave construction, a figured textile designer must be apt at evaluating the interaction of two or more weaves effects disposed across width and height of the textile's repeat. Through study and experimentation with two or more weaves in the same textile, a designer gains control over appearance and function to create figured fabrics that satisfy his or her objectives.

This chapter begins with a weave glossary that includes a discussion of weave, warp, and weft functions. The glossary is followed by a section that reviews criteria for choosing ground and pattern weaves. Building compound structures and weave compatibility are discussed. Examples are given that illustrate how binding systems may characterize a pattern or ground effect. Guidelines for structural evaluation of a figured textile conclude the chapter.

Weave glossary

Warp, Weft; End, Pick; Weave, Weave Repeat

Woven cloth is composed of two sets of threads that interlace at right angles to each other: warp and weft. The warp is held under tension and in parallel order by the loom, while the weft is inserted at right angles over and under the warp threads. A single thread of the warp is called a warp end, or end; a single insertion of a weft thread is called a weft pick, or pick. The order in which the ends interlace with picks is called the weave. One complete cycle of interlacement is the weave repeat.

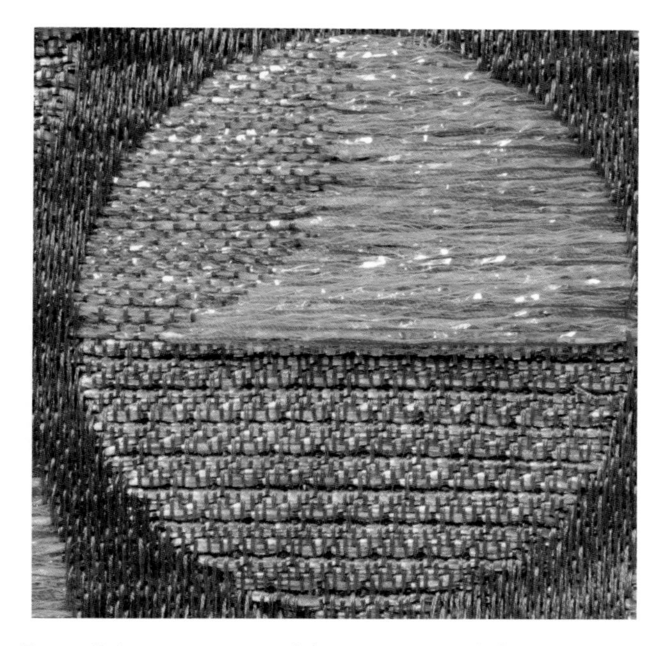

Figure 7.1. *Warp satin with long pattern weft floats:* Happles by Melanie Olde, LFS; *textile at real size; detail* above.

Simple Weaves, Compound Weaves

If all ends of a warp interlace according to a single interlacement order with one weft series, the resulting weave structure is called a simple weave.

If a group of ends or picks has a distinct function and/or follows a different interlacement order from other ends and picks, the resulting weave structure is called a compound or multiple weave.

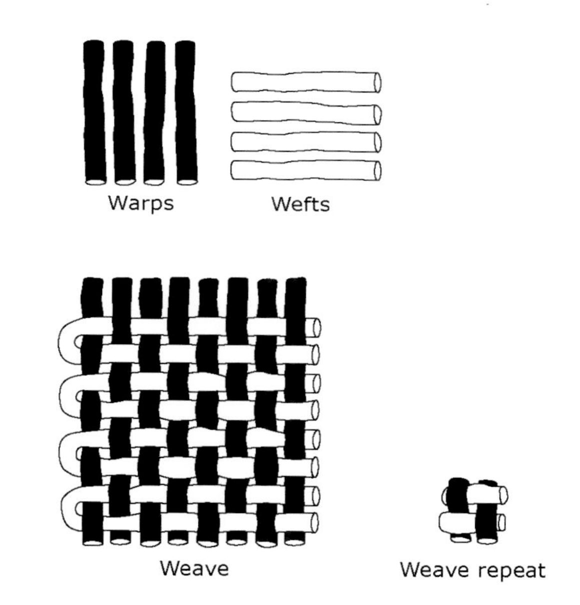

Figure 7.2. *Warp, weft, weave, and weave repeat.*

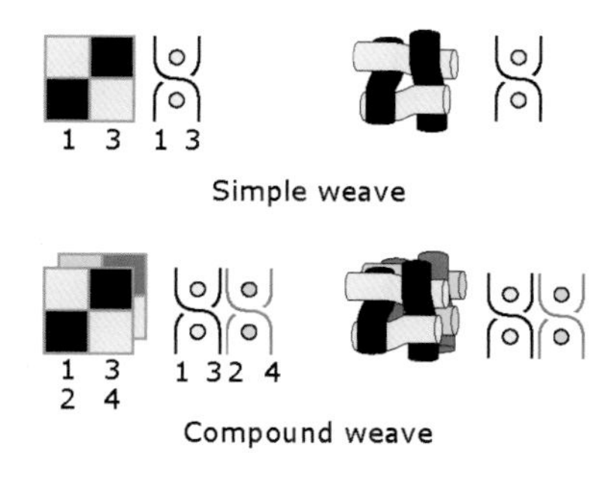

Figure 7.3. *Simple weave structure: plain weave. Compound weave structure: doublecloth, with two warp and two weft series.*

Weave Function

A *ground weave* is the supporting weave structure of a figured textile and is woven with a ground warp and weft series, and appears on the face of the cloth.

If the ground weave continues beneath patterning elements, the term used in this text is *foundation weave* or simply *foundation*. When a portion of the ground ends or picks of a textile are disengaged to produce patterning (see self-patterning warps or wefts—Chapter 6), and the ends and picks that remain below interlace to maintain the cloth's structure, this underlying structure may be referred to as the *bed* or foundation.

The term *pattern weave* may refer to simple or compound structures used to create patterning on the textile's surface.

The term *binding weave* generally refers to a simple weave that binds auxiliary or patterning elements within a compound structure, whereas *binding* is used for any form of interlacement.

In this text the term *binding points* refers to points that are added manually or by automatic functions of a software to the pointpaper or to single ends or picks in a weave structure so as to bind, or tack, overlong floats of weft or warp, stitch weave layers together, and so forth.

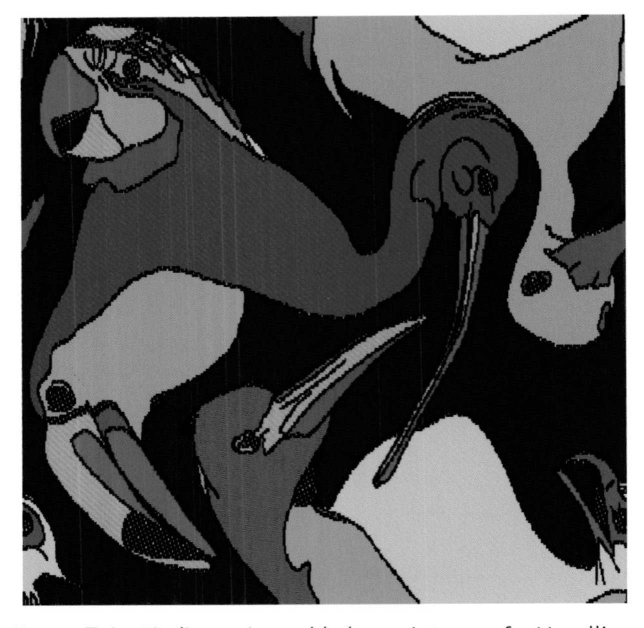

Figure 7.4. *Binding points added to pointpaper for* Uccelli.

Figure 7.5. *Plain weave with patterning warp and binding weft, LFCHT.*

Ground Weave and Ground Effect; Pattern Weave and Pattern Effect

There are times when the ground weave and ground effect do not coincide. Figure 7.5 shows a textile in which the ground weave creates the pattern, while a structure with a pattern warp on the face creates the ground effect. (See Design Glossary, Chapter 5.)

A *pattern weave* is any weave, simple or compound, that creates patterning. Pattern weave and *pattern effect* are used synonymously throughout this text.

Warp Series and Function

A group of warp ends with a distinct function and/or binding system is called a warp series.

The *ground warp* (also called foundation or main warp in many texts) interlaces with one or more series of wefts to construct the ground weave(s) of a figured textile. If any end of a ground warp is removed from the weave, a void is created in the textile's surface.

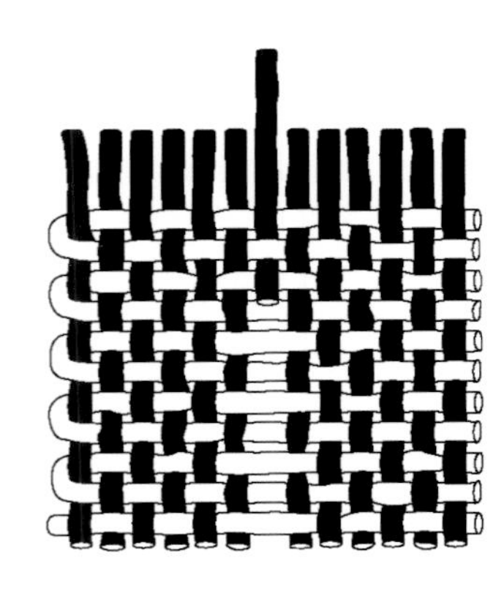

Figure 7.6. *Ground end removed from plain weave.*

A variation on a ground warp is an ***inner warp***, which functions as a ground or main warp to create the textile's basic structure, but does not appear on the surface, as in weft-faced compound tabby and twill. Yet another variation of ground warp function occurs when all or a portion of the ground warp ends float above the ground weft to create patterning floats on the textile's surface. When this occurs, the term ***self-patterning warp*** is employed. See Figure 7.7.

A ***pattern warp*** is any warp series that decorates the surface of the textile but does not contribute to the formation of the ground weave of a figured textile. If removed, a pattern end does not create an empty space in the ground structure, but does create a void in the patterning. A pattern warp may be continuous or discontinuous, occurring at regular or irregular intervals across the width of the textile.

A ***pile warp***, like a pattern warp, decorates the surface of a textile, but is generally used to create the loops or tufts of warp velvet. It does not contribute to the construction of the ground and may be continuous or discontinuous across the width of the textile. If one or more ends of the pile warp are removed, this will not create a void or weak point in the ground weave, but would create a void in the compact surface of the cut or uncut pile (removed pile not shown).

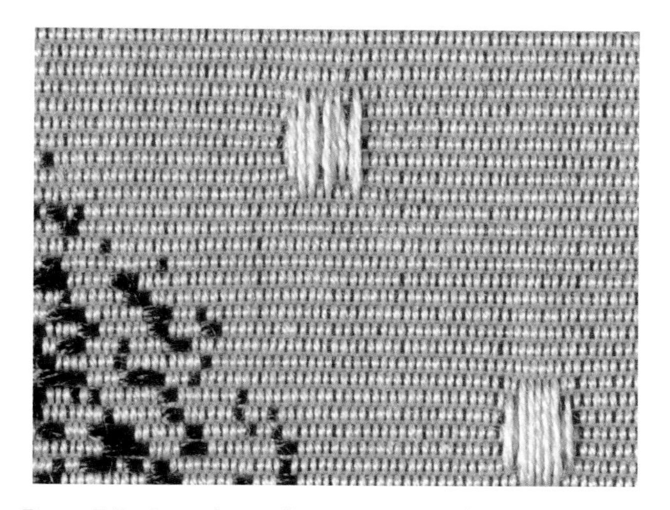

Figure 7.7. *Ground warp floats,* Fingerprint *by Anne-Birgette Hansen, LFS.*

Figure 7.8. *A textile with one ground and three pattern warps.*

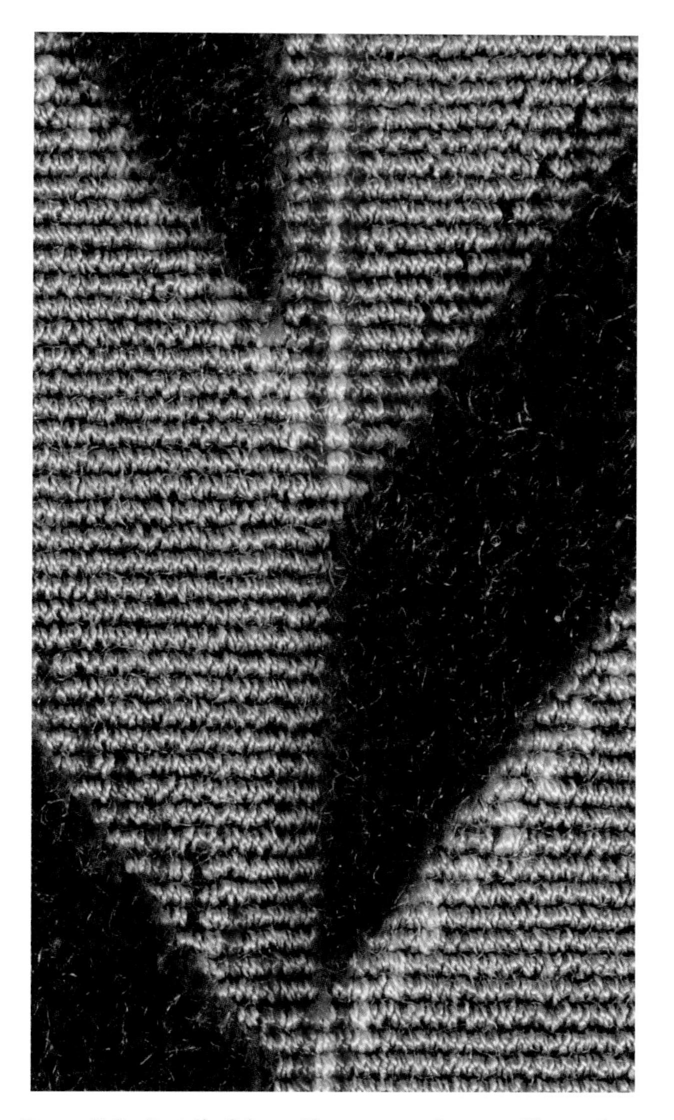

Figure 7.9. *Detail of the uniform cut and uncut effects of a ciselé velvet,* Chevron *by Berthe Forchammer, LFS.*

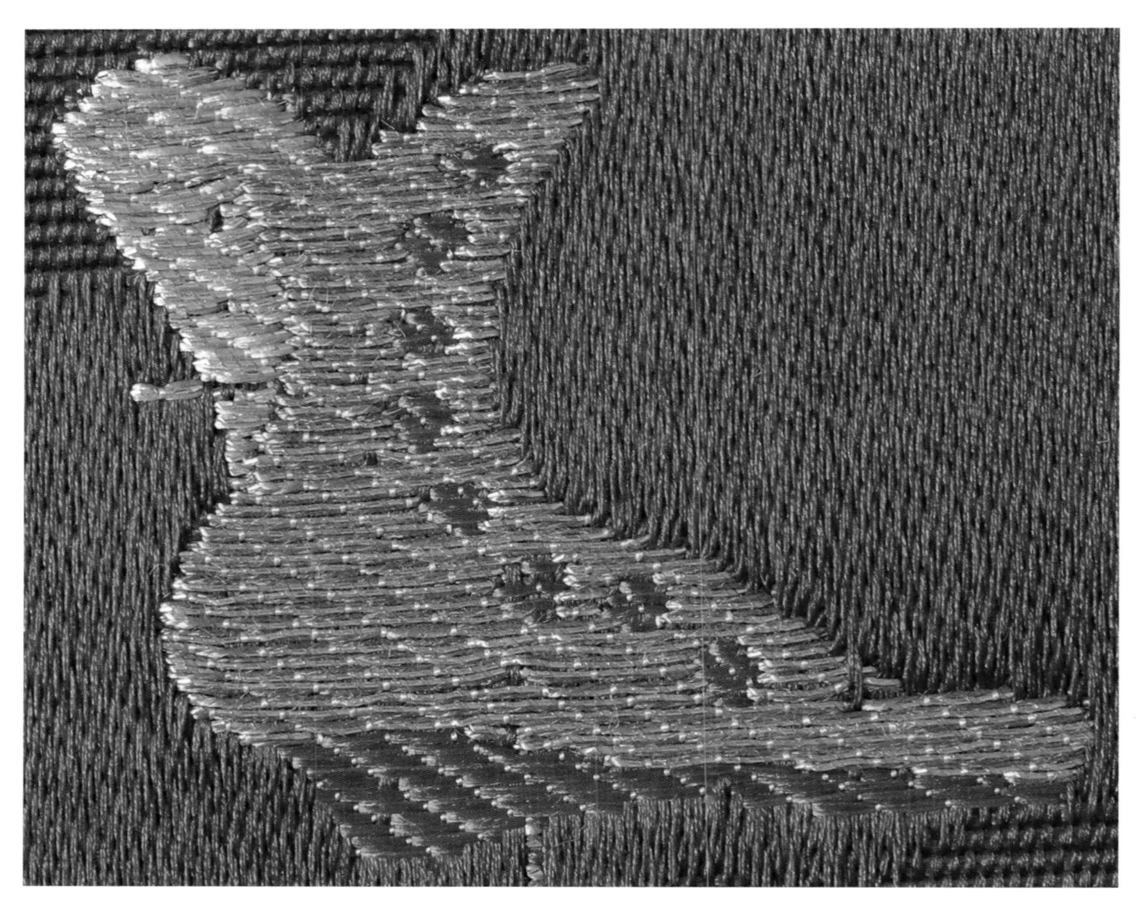

Figure 7.10. *The binding warp of a lampas with satin ground binds the gold and silk pattern wefts in 1/3 S twill.* Buttoned Boots *by Heidemarie Hohenbüechler, LFS.*

A ***binding warp*** is an auxiliary warp series that serves to bind either the complementary wefts of compound tabby or twill, or any of the pattern wefts defined later in this weave glossary. Removal of a binding end will create longer weft floats where these remain unbound.

Complementary warp: a complementary warp (not shown) has a double function in that it alternates with one or more equivalent warp series to create the face, reverse, and/or patterning of a figured textile. By itself, a complementary warp does not create the ground structure. Other equivalent series are needed to complete the ground weave.

Weft Series and Function

A series of picks with a distinct function and/or binding system is called a ***weft series.***

A ground weft, also called foundation or main weft, interlaces with one or more series of warps to construct the ground weave of a figured textile. If any pick of the ground weft is removed from the weave, a void is created on the textile's surface.

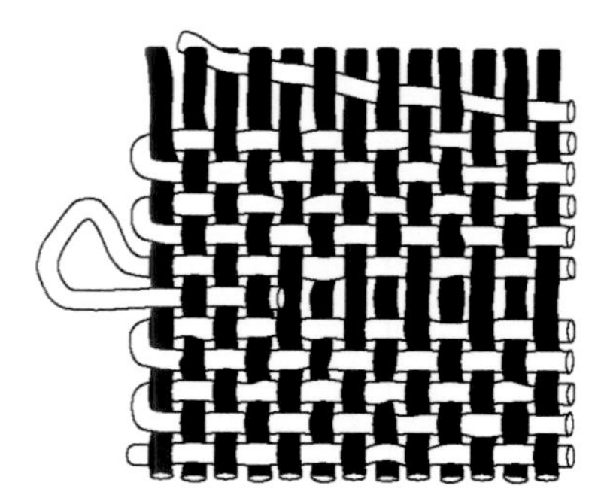

Figure 7.11. *Ground weft removed from plain weave.*

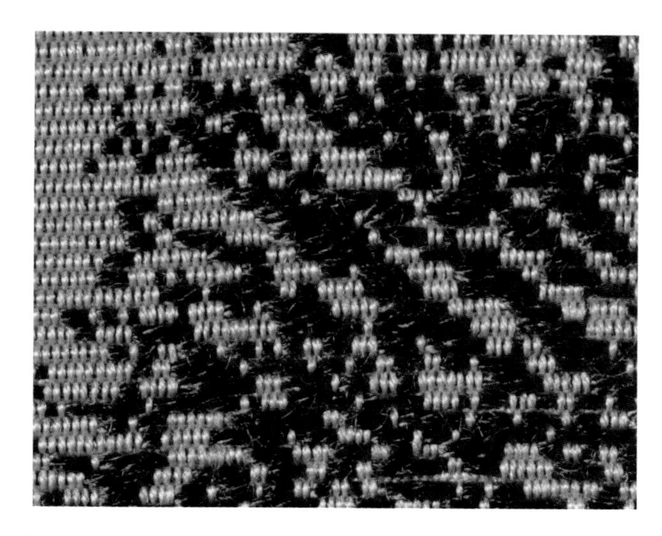

Figure 7.12. *Self-patterning weft effect,* Fingerprint *by Anna-Birgitte Hansen, LFS.*

A variation of ground weft function occurs when all or a portion of the ground weft picks are disengaged from the ground weave to create floats that decorate the surface of the textile. When this occurs, the term ***self-patterning weft*** is employed.

A ***pattern weft*** creates figuring on the textile's surface or rests on the reverse, but does not contribute to the formation of the ground weave. It travels across the warp from selvage to selvage and if removed does not create a void in the ground structure, but does produce an irregularity in the patterning. A pattern weft may be continuous or discontinuous along the height of a textile. Other commonly used terms for pattern weft are extra weft, supplementary weft, lancé weft, and figuring weft.

Figure 7.13. *Pattern weft removed from* Drago d'Oro *by Alberto Leoni, figured satin with pattern wefts, LFS.*

A ***brocade weft*** or brocading weft is a pattern weft used to weave a figure in a limited area across the textile's width and is not bound into the selvage. It may be continuous or discontinuous along the textile's height. If removed from the cloth, it does not create a void in the ground weave, but will create a void in the figuring.

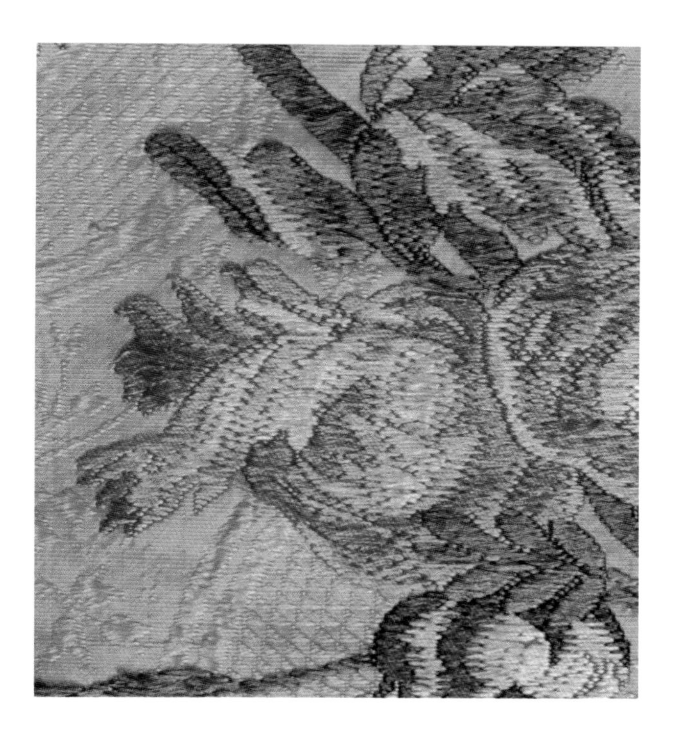

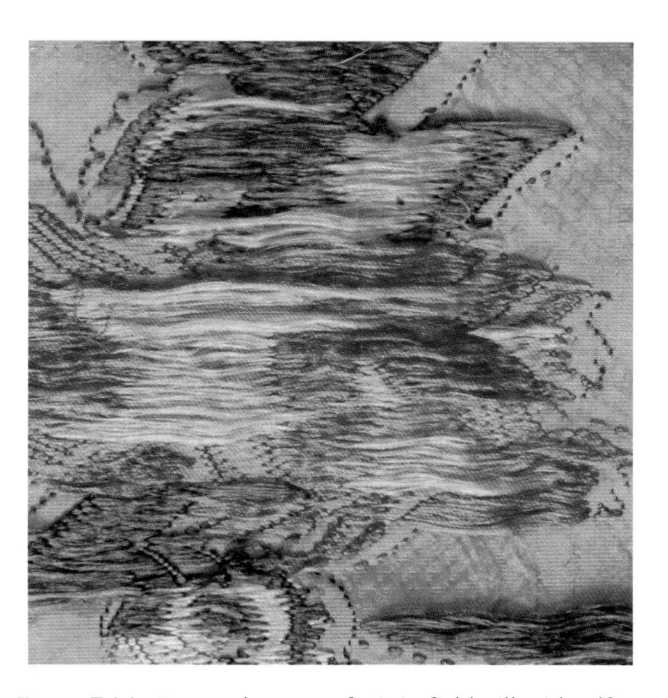

Figure 7.14. *Face and reverse of a Spitafields silk with self-patterning and multiple brocading wefts. Courtesy of Cora Ginsburg LLC.*

A *binding weft* is an auxiliary weft series that serves to bind either the complementary warps of compound tabby or twill, or any of the pattern warps defined in this section.

Complementary weft: a complementary weft has multiple functions in that it alternates with one or more equivalent weft series to create the face, reverse, and/or patterning of a figured textile. By itself, a complementary weft does not create the ground structure. Together with other equivalent series, it "completes" the ground structures. Example: weft-faced compound tabby or twill, or weft double-faced cloth.

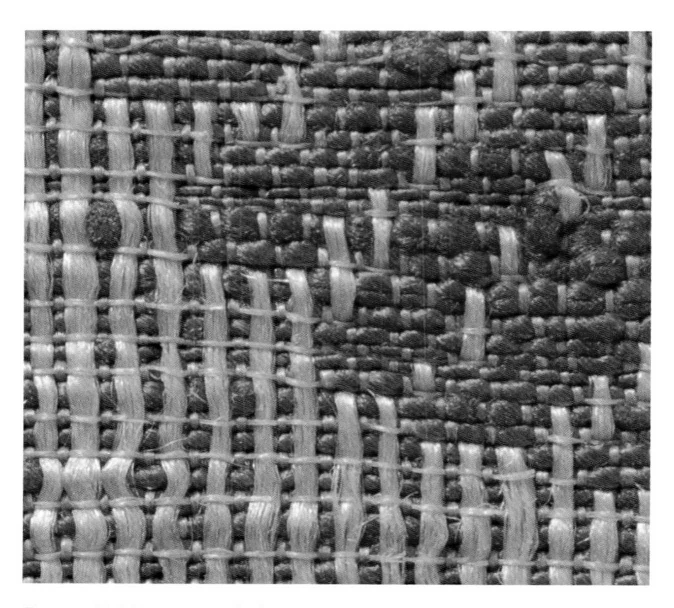

Figure 7.15. *Figured plain weave with pattern warp bound by binding weft, LFCHT.*

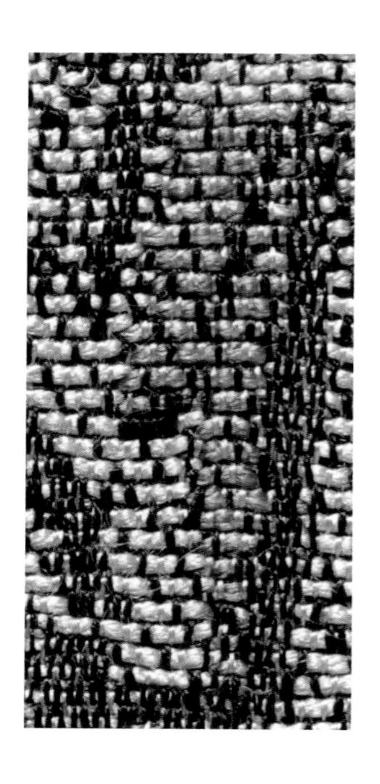

Figure 7.16. Left: Sheetal's Leaves *woven with two complementary wefts, by Sheetal Khanna-Ravish, LFS.*

Choosing weaves

When selecting weaves for a figured textile, what considerations influence weave choice? Aesthetic or structural? This depends on an array of factors tied to a textile's purpose. In this book, priority is given to the desired appearance of the cloth, after which correct structural solutions or compromises are sought with the surface in mind.

The following guidelines are based on the assumption that weaves should be stable and designed to support a reasonable amount of stress. For those weaving works of art or textiles with specific properties, it is up to the artist or designer to adapt these indications to the particular parameters of their work.

Factors that Affect All Weaves in a Figured Textile

Three factors influence all weaves in a given textile, irrespective of function and technique: the loom setup, warp sett, and warp materials.

Loom setup: every figuring device raises a maximum number of independently controlled ends. Most industrial and many handlooms are "harnessed" or set up to repeat this maximum number several times across the width of the loom. The total number of independent ends limits the choice of weave structures that may be employed for weaves that "travel" over the border of a repeating design. In practice, the warp repeat of all such weaves must be a submultiple of the maximum number of individually controlled ends. Example: if a loom's patterning capacity is 1,728 ends, then only weaves with a warp repeat that divides this number evenly may be used, such as 2, 4, 8, 12, 16, 32, 36, 64, 96, and so forth. Weaves having a warp repeat of five or multiples of five would produce a defect at the edge of the repeat, if used with this loom setup.

Warp sett: there is usually little or no margin to modify warp sett, as this depends on the spacing in the comber board of the harness cords that control figuring ends. A new generation of single-end control looms allows for some modification, but to date, varying the sett of most looms remains time consuming. See Chapter 4.

Warp sett affects weft float length. Example: 100 ends per centimeter / 254 ends per inch, a 32-end weft satin produces a float of 0.32 centimeters in length, or 0.08 inches. At 25 ends per centimeter / 64 ends per inch, the same weave produces a 1.28 centimeter / 0.32 inch float.

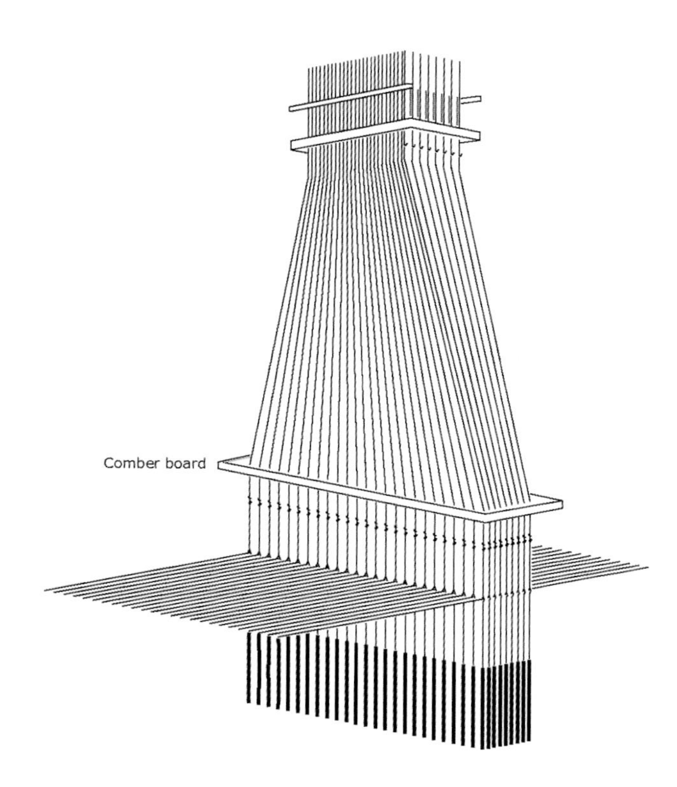

Figure 7.17. *Harness with comber board, also called harness reed.*

Fiber content and yarn construction of ***warp materials*** affect warp elasticity and capacity to compensate for take-up differences. "S" or "Z" twist of ground warp yarns determine the choice between "S" or "Z" twills or warp satins as ground weaves; yarn count and construction are indicators of the coverage a yarn can provide and thus determine the choice between tighter or looser weave structures, as well as between warp-faced, weft-faced, or balanced weaves. See Figure 7.19.

Weave Choice Based on Function and Appearance

Ground weaves provide the supporting structure of a multi-effect textile, and like the foundation of a building, determine its stability and capacity to support the stresses to which a construction is subjected. Based on warp sett, and warp and weft materials, the weaver chooses a weave with sufficient interlacement to guarantee a solid construction.

The ground weave has an important impact on the overall appearance, texture, and hand of the textile. Each of the three weave families lends a distinct character to the surface.

- **Plain weave and derivatives—grainy, tight**
- **Twill—graphic, with an apparent diagonal line, stable but flexible**
- **Satin—smooth, material enhancing, flexible**

Warp/weft materials and color affect the choice between balanced, warp, and weft-faced weaves. Technique (figured plain weave, damask, brocade, lampas, velvet, etc.) influences weave choice. Example: taqueté, samitum, and weft brocade are woven with weft-faced weaves. See Chapter 6.

Weave structures chosen for the ground determine the weft sett for the entire textile and limit the selection of weaves for binding extra warp and weft series on the reverse, as these must be compatible with the ground structure. Ultimately, all other weaves in a textile are affected by this fundamental choice.

S

Z

Warp satin 8
shift 5

Warp satin 8
shift 3

Figure 7.18. *Yarn twist and ply determine choice between S or Z satin.*

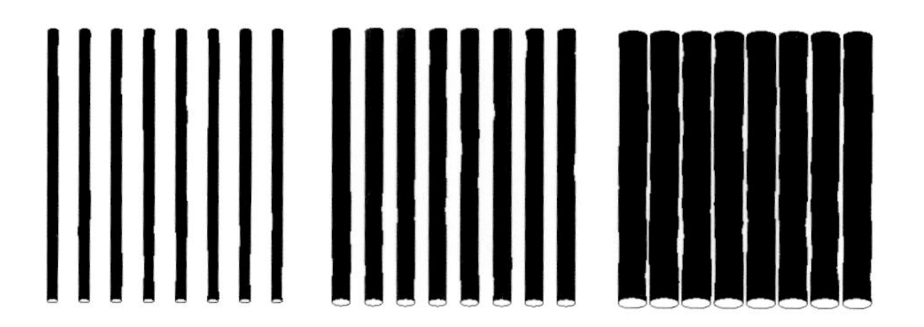

Figure 7.19. *The combination of warp material and sett determines coverage.*

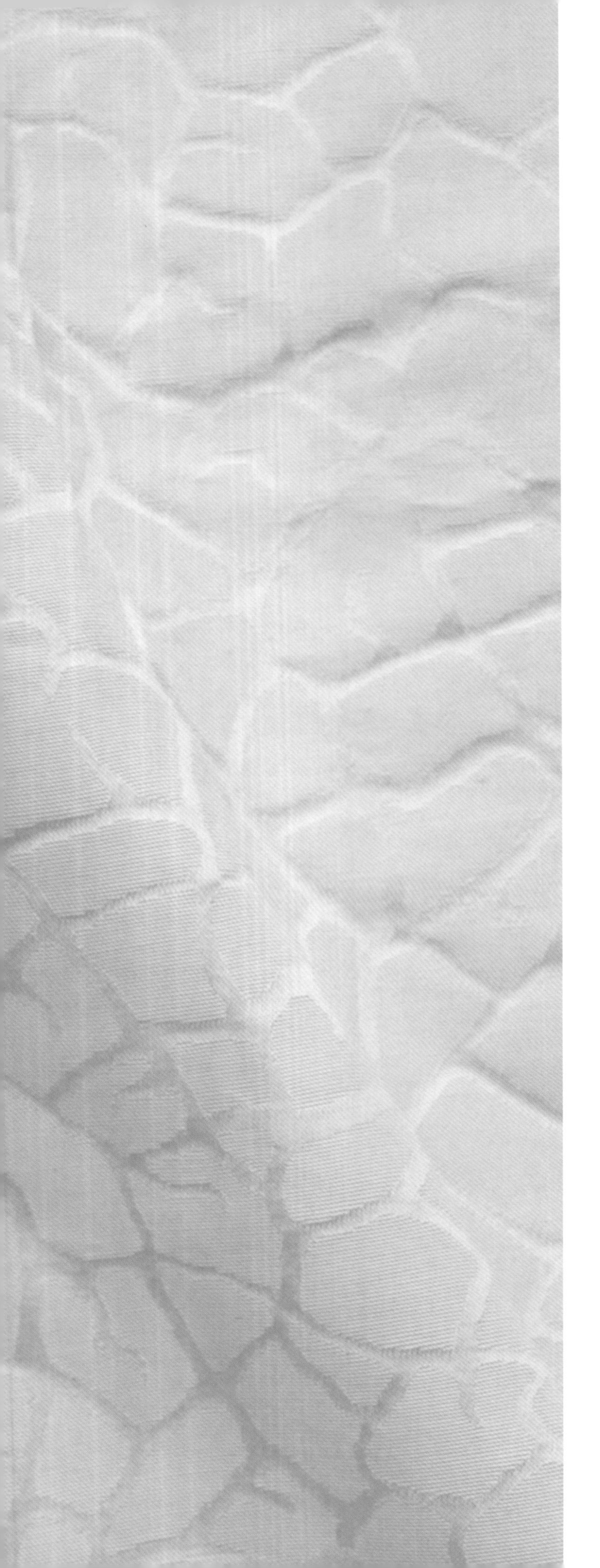

To create a visible figure, a ***pattern weave*** must contrast with the ground structure. In a figured textile with only one warp and weft series, contrast depends on differences in warp and weft materials, weave class, repeat size, float length, or a shift from warp to weft face or the reverse. Any of these factors produces some degree of contrast between patterning and ground weave, as illustrated in Chapter 6, *Damask*.

Building weaves

Patterning with Compound Structures

Many figuring techniques employ more than one warp or weft series to create patterning. Techniques shown in Chapter 6 include: self-patterning warps/wefts, extra warps/wefts, exchange of equivalent surfaces (double-faced or complementary structures), and exchange of equal or unequal layers (multiple cloths). Often Jacquard textiles combine several techniques, each with its own possibilities and challenges.

Weave Compatibility

When building weaves with more than one warp or weft series, a designer seeks solutions that maintain the desired appearance of the surface weave and that respect the technical parameters of a given technique, warp setup, and materials. For weaves to be compatible, the binding points of one weave must not interfere with those of the other weaves in a compound structure, as this can modify the appearance of the surface and/or add take-up to one or more ends in the weave repeat. Solutions vary for warp and weft-faced structures, as illustrated in the figures that follow.

Compound Warp Structures, Compound Weft Structures

In the drafts for double-faced warp satin in Figure 7.21, both weaves belong to the same class, have the same warp/weft repeat, and use the same shift. The binding points of the lower weave fall between the long warp floats of the upper weave, and thus are perfectly aligned and invisible. Both faces will appear as warp satin, and no unwonted or unequal take-up is generated.

Figure 7.20. Leaf Skeleton *by Janina von Weissenberg, FLS. Ground in 2/2 basket, patterning woven with several shaded satins.*

As in the previous compound structure, both weaves of the double-faced twill in Figure 7.22 belong to the same class and have the same repeat. The twills' diagonals match. The binding points of the warp-faced twill below are aligned to fall between the longer weft floats of the upper weave, and again the weaves are perfectly compatible.

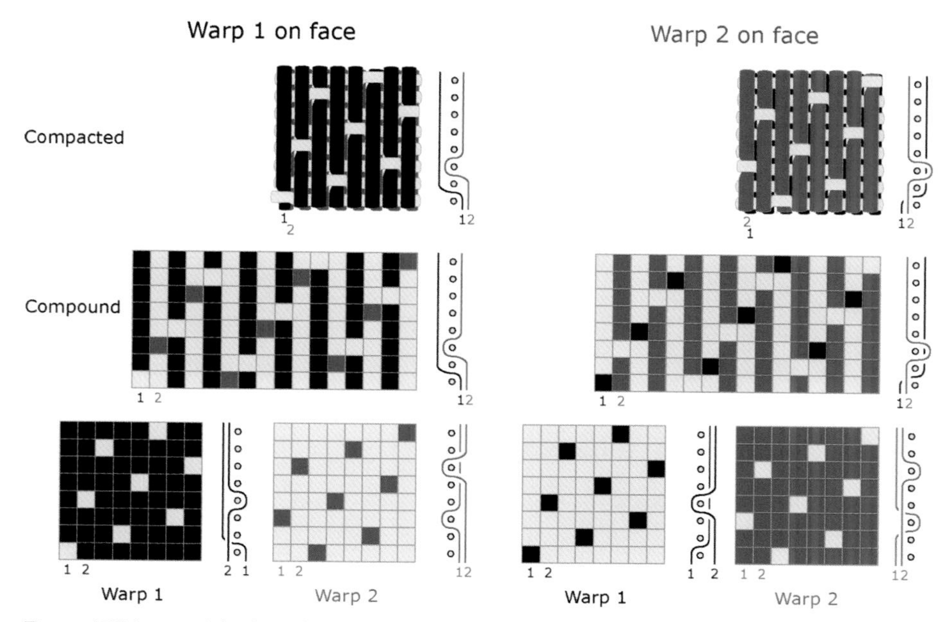

Figure 7.21. *Double-faced warp satin 8.*

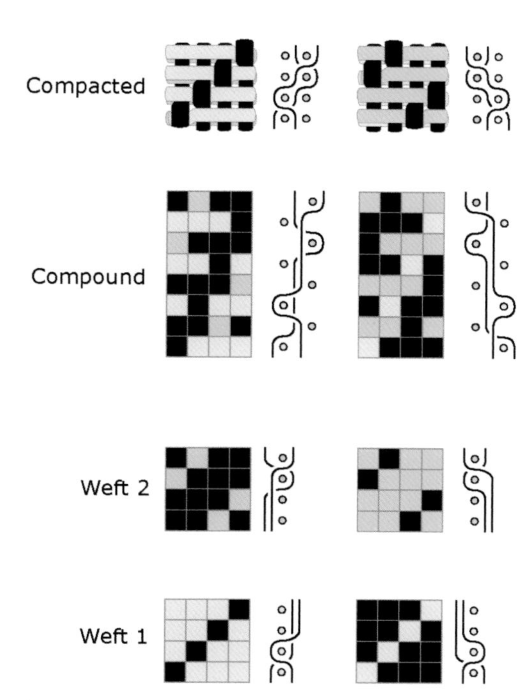

Figure 7.22. *Double-faced 1/3 weft twill.*

Patterning with Extra Wefts

Patterning with extra wefts opens a world of possibilities. Weave class, float length, and color changes are no longer subject to many of the structural limitations that restrict choices for ground wefts, as a pattern weft does not contribute to building the ground weaves of a textile. Weft patterning is also free of the limitations intrinsic in warp patterning, which is always restricted by take-up considerations. Because weft patterning is more versatile than warp patterning, it is more frequently used in figured textile production; for these reasons discussion focuses on weft patterning rather than warp patterning.

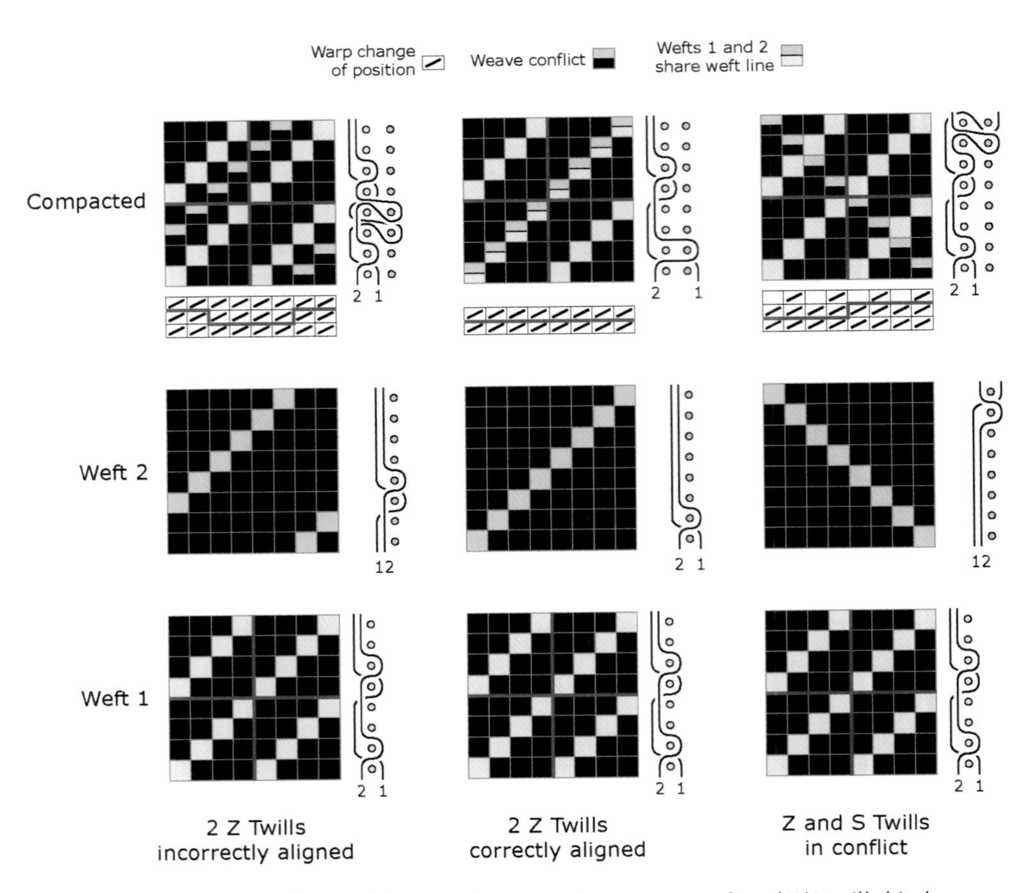

Figure 7.23. *Warp-faced 3/1 twills create the ground weave; warp-faced 7/1 twills bind a pattern weft on the reverse.*

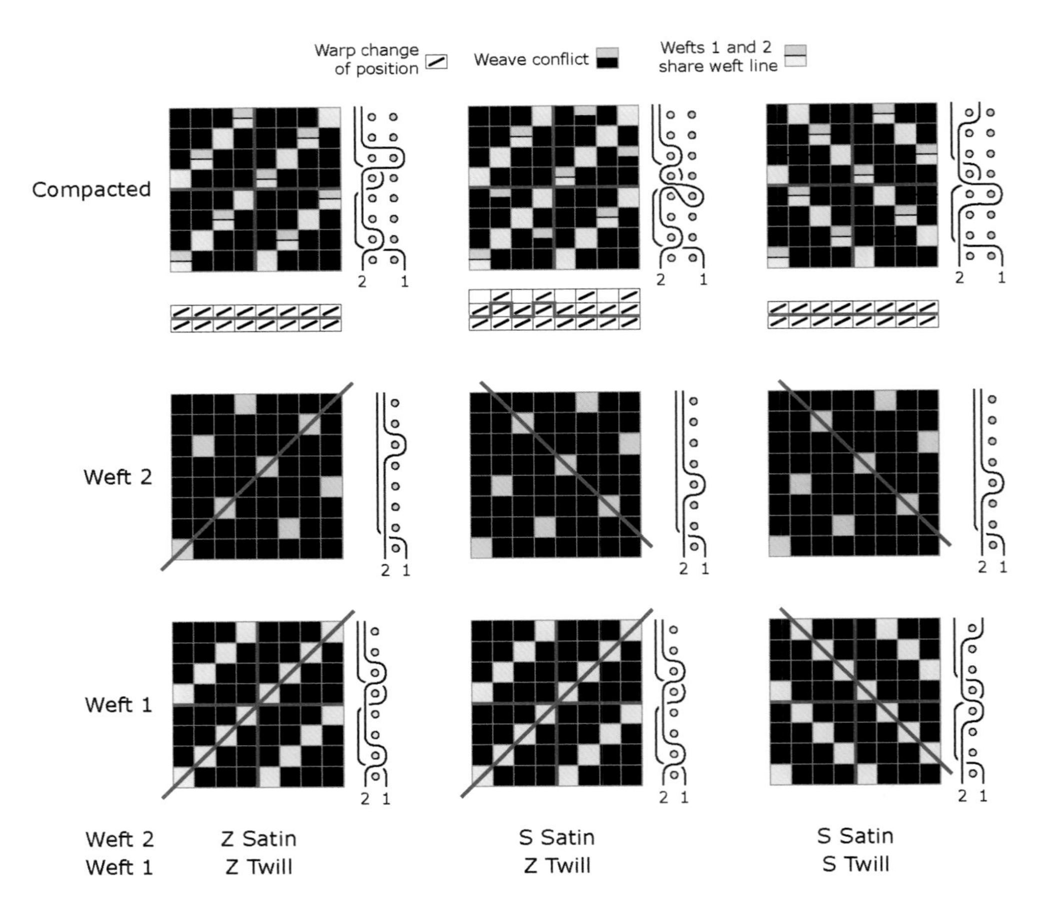

Figure 7.24. *Warp-faced 3/1 twills create the ground weave; warp-faced satin 8 binds a pattern weft on the reverse.*

Binding Extra Wefts on the Reverse of Warp-faced Weaves

When binding extra wefts on the reverse of a warp-faced weave, the binding points of the weft that lies below must coincide with the lowered ends of the upper weave, otherwise these will "cut" the warp floats that lie above and alter the appearance of the ground weave, and may add take-up to a portion or all of the warp ends. Figures 7.23 and 7.24 illustrate the results of both compatible and incorrect solutions for binding wefts beneath warp-faced weaves. Incorrectly aligned but potentially compatible; correctly aligned, compatible weaves; and incompatible S and Z combinations of otherwise compatible weaves are shown.

If two weaves are compatible, the combination of their binding systems does not increase the take-up of any end in the weave repeat. In Figures 7.23 and 7.24, the number of exchanges from face to reverse of every end in each compound weave is indicated by a slash on squared paper below the compacted draft. A horizontal red line separates the slash/exchanges of the first eight picks of the compacted draft (below the red line) from those of the second eight picks of the draft (above the red line). If the single weaves are compatible, and no take-up is added to any one end, the total number of slashes is equal for every end of the compound structure.

Some generalizations regarding warp-faced compound structures with a single warp system follow: 1) the weaves used to bind pattern wefts on the reverse of a warp-faced ground weave must also be warp-faced, 2) all binding points of a weave used to bind a pattern weft on the reverse must coincide with those of the uppermost weave, 3) the weave that binds the pattern weft on the reverse must be a multiple of the ground weave, 4) if twills and satins are used in combination, both weaves must share the same diagonal, S or Z.

Figure 7.25. Gondola II *by Tina Moor, LFS. A ground weft is bound in 3/1 twill, while a warp satin 16 binds a pattern weft on the reverse.*

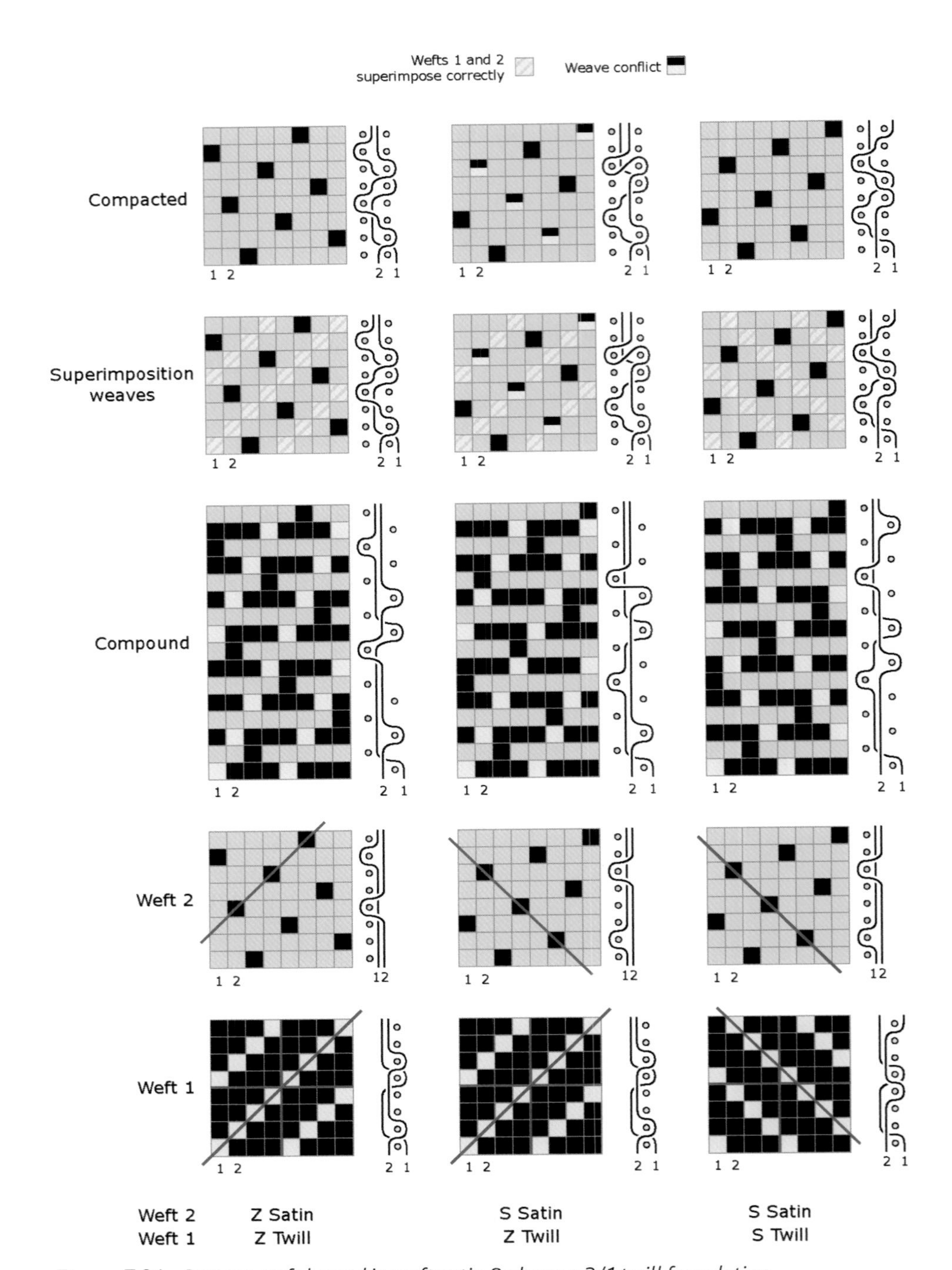

Figure 7.26. *Pattern weft bound in weft satin 8 above a 3/1 twill foundation.*

Patterning on the Face with Extra Wefts

Figure 7.26 shows a commonly used combination of weaves for weft patterning: satins and twills. As in the preceding cases, the two simple weaves are potentially compatible. Defects in the compound weave at the center of the figure are due to an incorrect alignment of the S / Z diagonal of the two structures. Where the diagonals of the twill and satin match, the weaves are compatible Note how the binding points of the lower weave "hide" between the weft floats of the upper weave. As the patterning picks are beaten in, these are free to ride over the ground picks without impediment. The surface of compatible weaves is uniform, and no increase in take-up occurs.

When a pattern weft is bound on the face of a compound structure having a single warp system, 1) the weft-faced weave must be a multiple of the foundation weave, 2) the binding points of the weft-faced weave must coincide with raised ends of the foundation weave, 3) if twills or satins are used in combinations, the diagonals of both weaves must match, and 4) the binding points (lowered ends) of the foundation weave will be aligned between the floats of the pattern weft.

Figure 7.27. *Weft-faced satins of various lengths bind weft-patterning above a damask ground.* Magnolia *by Masako Hosono, LFS.*

Other Methods for Binding Pattern Wefts on the Face

Not all weft-patterning is bound by regular weaves. Overly long floats may be bound by adding points to the pointpaper design to create textures, add details or lines to a motif, or create small patterning. Such devices can be employed when designing handwoven or industrially woven textiles.

On the next two pages:

Figure 7.28. Left page: *Figured textile with several weft-patterning effects. Binding points were added by hand to the pointpaper to reduce float length and further pattern the ground and motifs.* Pine Motif *by Tuulia Lampinen, LFS.*

Figure 7.29. Right page: *All patterning in* Green Lancé *is created by weft floats with no other binding than details within the motifs; Francesca Leoni, LFS. In weft-patterned* Yellow Roses *by Francesco Sala, LFS, binding points drawn by hand on the pointpaper add detail to the motifs.*

The Impact of Warp and Weft Sett on the Appearance of Weaves

We tend to think of plain and basket weaves as resembling the alternating squares of a checkerboard, we assume that regular twills produce a forty-five degree diagonal line, and we prefer satins with one shift rather than another without considering the impact of sett on the appearance of the weaves we choose. The ratio between warp and weft setts affects the appearance of all weaves. Often in Jacquard weaving, this ratio is not 1 to 1. See *aspect ratio*, Chapter 5.

If the designer wishes to maintain the square aspect of a plain weave or 2/2 basket, the forty-five degree diagonal of a twill, or the appearance of a specific satin, the ratio of warp to weft sett is taken into account and the aspect ratio calculated; only then are weaves chosen or designed based on the real dimensions that the interlacement of ends and picks occupies on the surface of the textile.

The ratio between warp and weft setts of the handwoven brocade in Figure 7.32 is unequal. When this brocade was designed, the maker had a full understanding of the impact of sett on weave appearance, as demonstrated by the quality of the patterning.

1:1 1:2 2:1

Figure 7.30. Above: *2/2 basket, 1/7 twill, and weft satin 8 as these appear when woven with a width to height aspect ratio, from left to right, of 1:1, 1:2, and 2:1.*

Figure 7.31. Right: *The weft-patterned motifs of* Grenade *were created by designing small geometric patterns on a pointpaper that matched the aspect ratio of the overall textile. Figured damask with weft-patterning, by Emelia Haglund, LFS.*

Figure 7.32. Facing page: *Seventeenth-century handwoven silk with gold brocading. Courtesy of Cora Ginsburg LLC.*

Figure 7.33. Facing page: *Detail, pointpaper, and image at real size of reconstructed eighteenth-century droguet, by Martin Ciszuk, FLS.* Above: *photograph of original sample. Courtesy of Martin Ciszuk.*

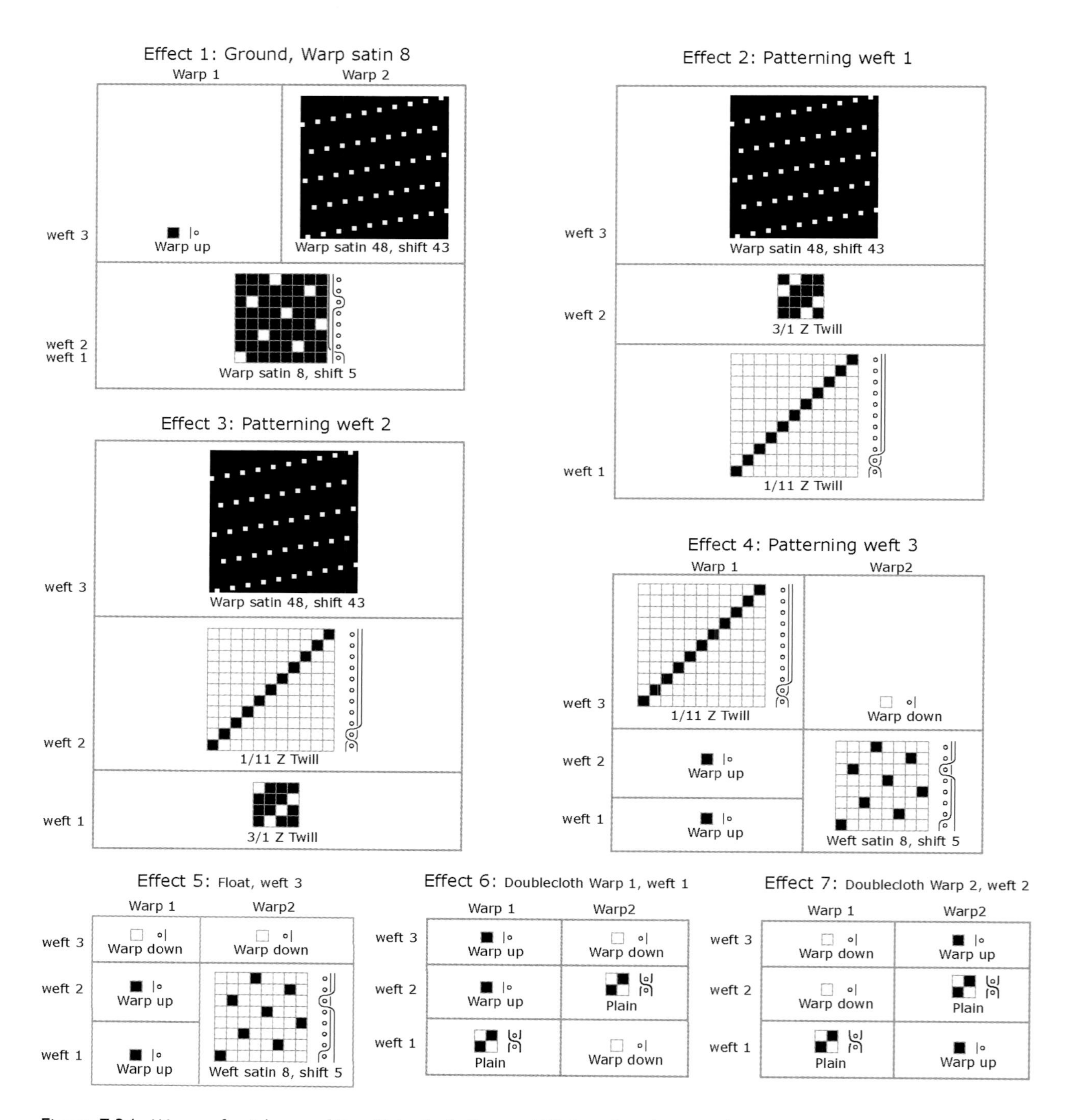

Figure 7.34. *Weaves for* Adam and Eve III, *by Rudy Kovacs, LFS. Detail and image of textile at real size.*

Custom-designed Weaves

To replicate the warp and weft patterning of an eighteenth-century droguet, using an industrial loom with a single warp series, the designer redrew each detail of the motifs to dimension on a pointpaper scaled to reflect the aspect ratio of the new textile. He then designed all the weaves and floats based on the warp parameters and weft sett of his modern replica.

Adam and Eve III combines several patterning techniques and all weave families. There are areas that weave in a single layer as warp satin, others as doublecloth, while two self-patterning wefts and one pattern weft create figuring on the face, float between layers, or are bound on the reverse. With shifting densities of ground and figuring structures, this project required careful technical evaluation to ensure that it would weave correctly.

Structural evaluation

Having first selected weaves for visual effect, a designer evaluates each weave structure in relation to the design and end use of the cloth, after which minimal to substantial changes are made as needed. Factors to consider are:

Take-up
Weave stability
Surface strength

Take-up

Take-up is the greatest limiting factor in Jacquard weaving. The single ends that compose a warp series must maintain an equal tension to ensure a clean shed and avoid warp breakage. For equal tension to be maintained, the amount of interlacement of each end must be approximately equal to the interlacement of every other end in a warp series wound together on the same beam.

To compare take-up between two weaves, the total number of binding points along each warp end is counted for an equal number of picks. If patterning wefts are present, but correctly bound on face or reverse, take-up remains unaffected. See Figure 7.36, Page 150.

Jacquard pattern "Afterparty"/2011

Designed by: Pirita Lauri, second year student in BA, Textile Arts and Design
Designed at: Aalto University, School of Arts, Design and Architecture (Helsinki, Finland)

About the pattern

Sketch for this jacquard pattern was made originally to be a print. Pattern has been rotated 90 degrees to be able to weave multicolored stripes in weft direction. Four different weft materials are alternated by the weaver in a free way. Vertical gaps between the floats are plain weave in order to avoid long floats in warp and to maintain equal tightness throughout the fabric.

Technical information

Loom: TC-1 (4moduls)
Warp material: SE
Density of the warp: 24y/cm
Weft materials: CO/PAN
Density of the weft: 19y/cm
Size of pattern file: 18cm x 17,17cm
Size of the woven pattern: aprox. 18cm x 21,5cm

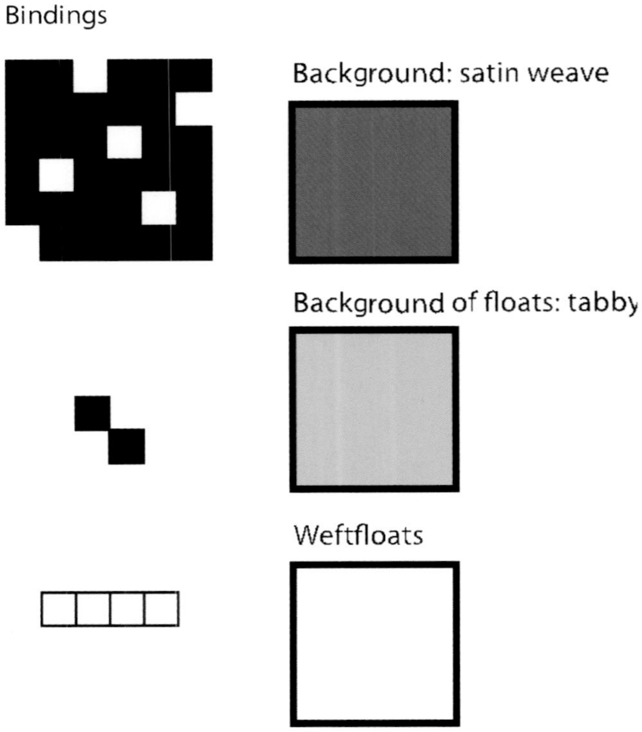

Bindings

Background: satin weave

Background of floats: tabby

Weftfloats

Figure 7.35. Afterparty *by Pirita Lauri, Aalto University Photographs and drawings courtesy of PL*

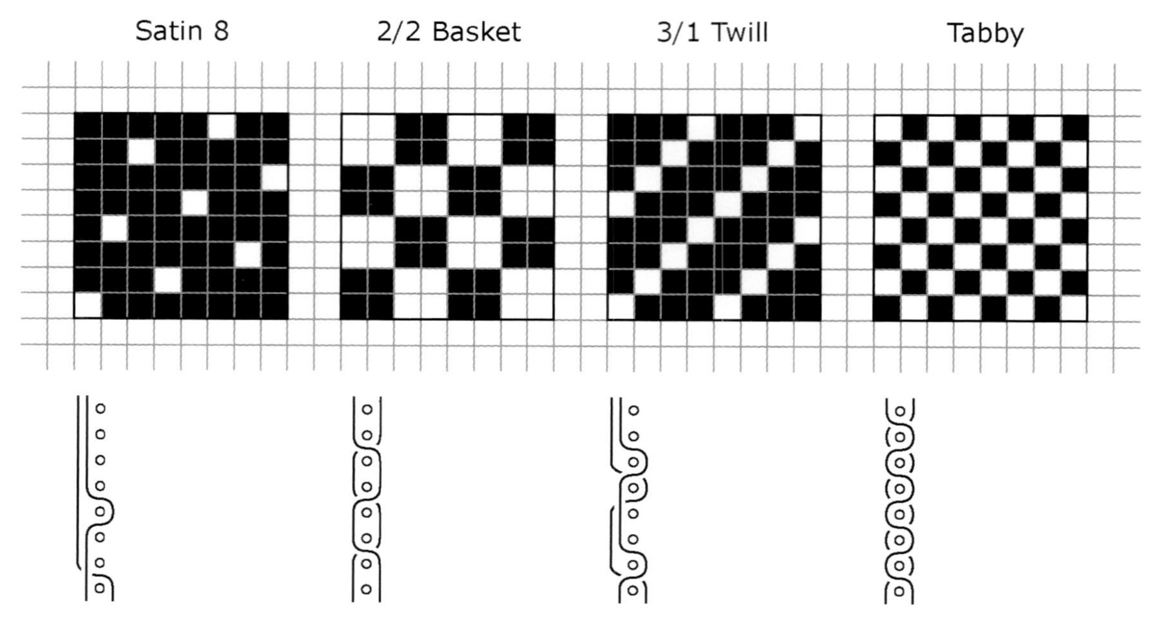

Figure 7.36. *Comparison of take-up of an 8-end satin, 2/2 basket, 3/1 twill, and plain weave.*

In Figure 7.36, the take-up of four commonly used weave structures can be compared. To facilitate comparison, eight picks are drafted, a number equal to the picks of the longest weave repeat shown, an 8-end satin on the left. From left to right, the number of binding points per weave in eight picks is: 1, 2, 2, and 4.

Differences in take-up between weaves in the same textile may or may not produce significant tension problems, depending on a number of factors, including warp elasticity: nonelastic fibers such as linen allow for little tension difference between ends and weave areas, whereas continuous filament yarns, such as silk organzine, will compensate for a larger degree of difference.

The distribution of weave effects in a design can cause or mitigate tension problems. Imagine a vertically striped textile woven in warp satin 8 and plain weave: no amount of warp elasticity can compensate for the different take-up of these two structures. If the same satin 8 and plain weave are used in combination to weave a checkerboard design, the vertical exchange between tighter and looser structures compensates for overall take-up differences, when woven on an elastic warp.

Two sets of drawings illustrate how design can affect take-up. Figure 7.37 shows a distribution of motifs along a vertical axis, which is more likely to generate problems if weaves with unequal take-up are selected. Figure 7.38 shows a distribution of motifs that compensates for differences in weave take-up.

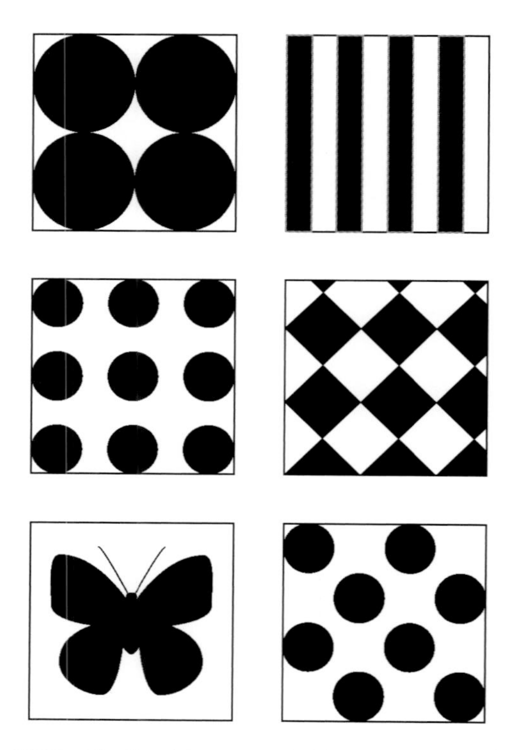

Figure 7.37. *Motifs with a potential for generating take-up differences.*

Figure 7.38. Facing page: Uccelli *by Daniela Tirabasso, FLS. The ten weaves of* Uccelli *(see relative reading note in Chapter 9 and pointpaper in Figure 7.4) generate substantially different take-ups, but distribution of weave areas compensates for any difference in warp shortening.* Inset: *problem-free distribution of motifs.*

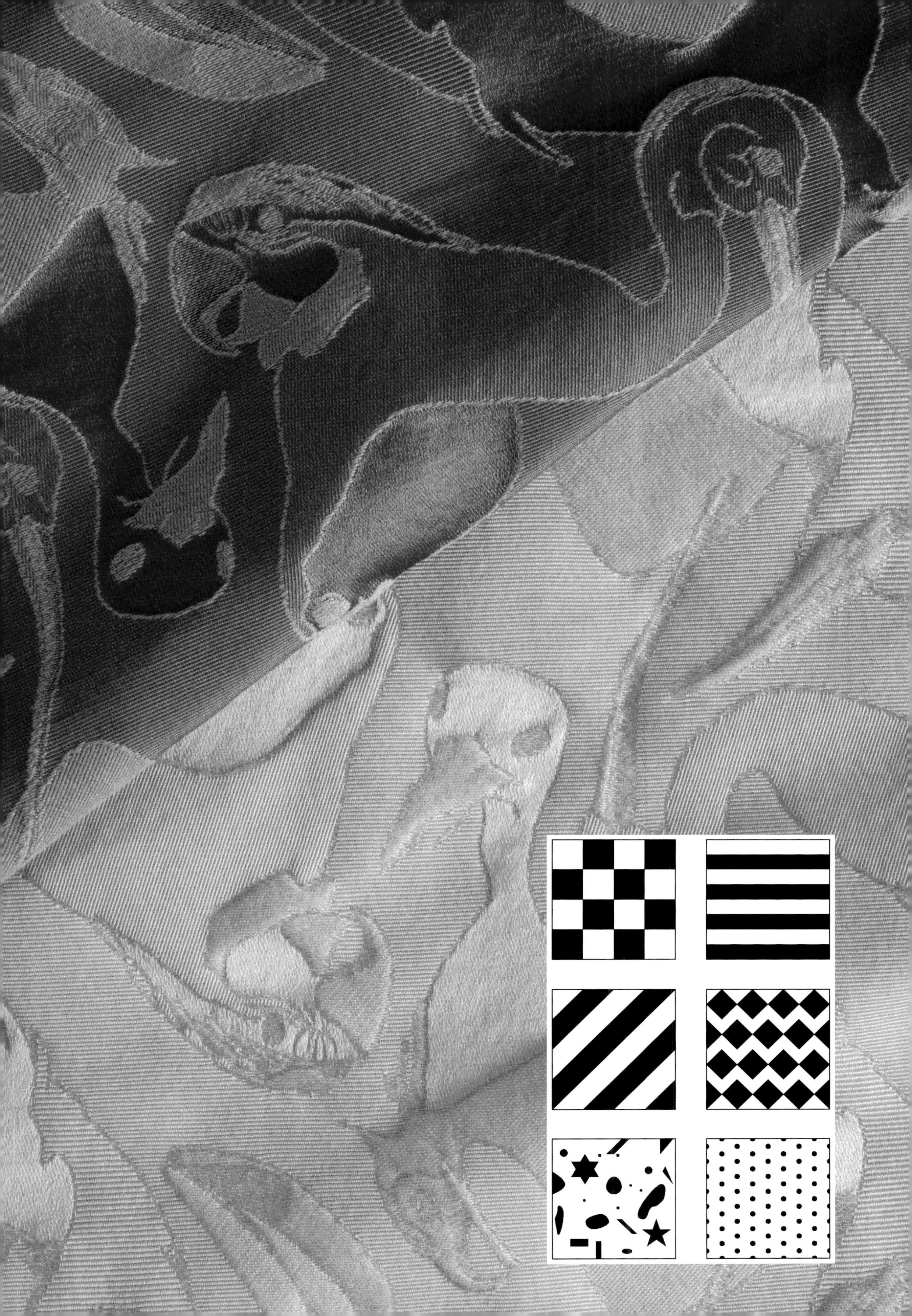

Weave Stability

Not all textiles, such as wall hangings, will be exposed to movement or "stresses" that require a weave to be stable. See Figure 7.1, *Happles,* and *Afterparty* in Figure 7.35. Evaluation of weave stability is based on a textile's end use. The stability of a weave depends on frequency and tightness of interlacement. In Figure 7.39, the weave on the right of each pair guarantees more stability, while maintaining similarity of appearance, warp/weft balance, and take-up.

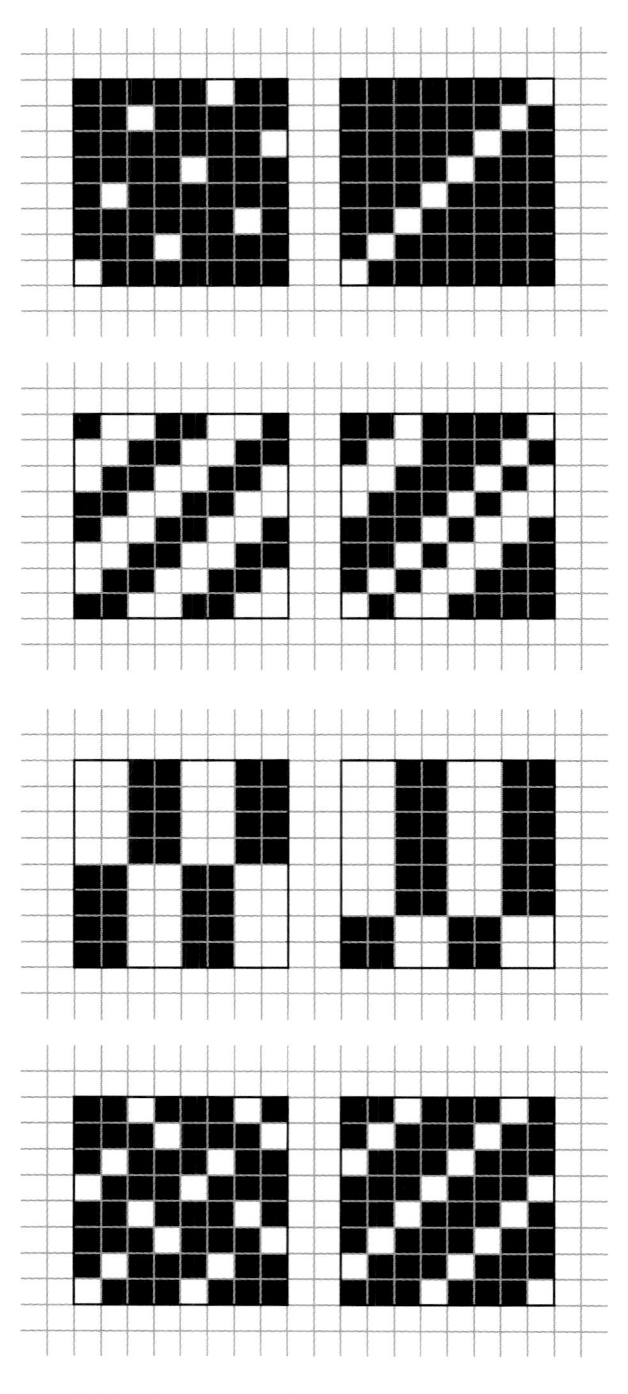

Figure 7.39. Above: *Pairs of weaves with the more stable structure shown on the right.*

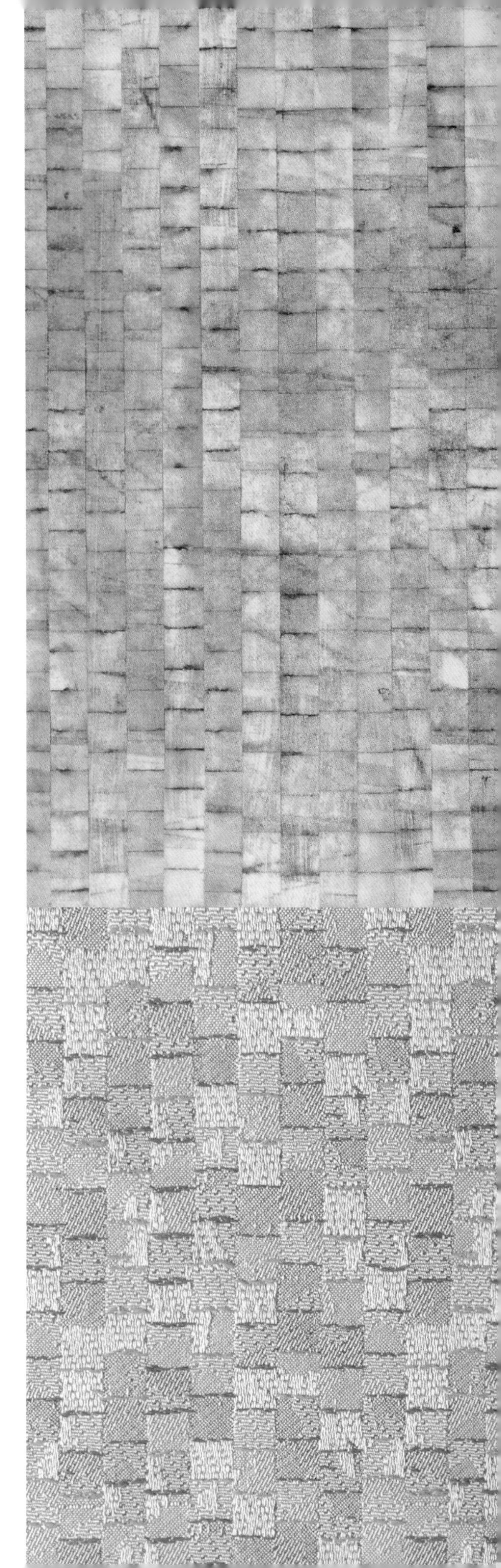

Surface Strength

As with weave stability, evaluation of the surface strength of a textile—its vulnerability to abrasion, snagging, and wear—is related to intended use. Factors that affect surface strength are:

Float length
Yarn construction and materials
Warp or weft-faced weave structure
Dimension of weave areas
Tightness of weave effects adjacent to areas with looser interlacement

In *Small Squares* the distribution and frequent exchange between tighter and looser weave structures ensures stability and equalizes the differences in take-up of the single weave effects.

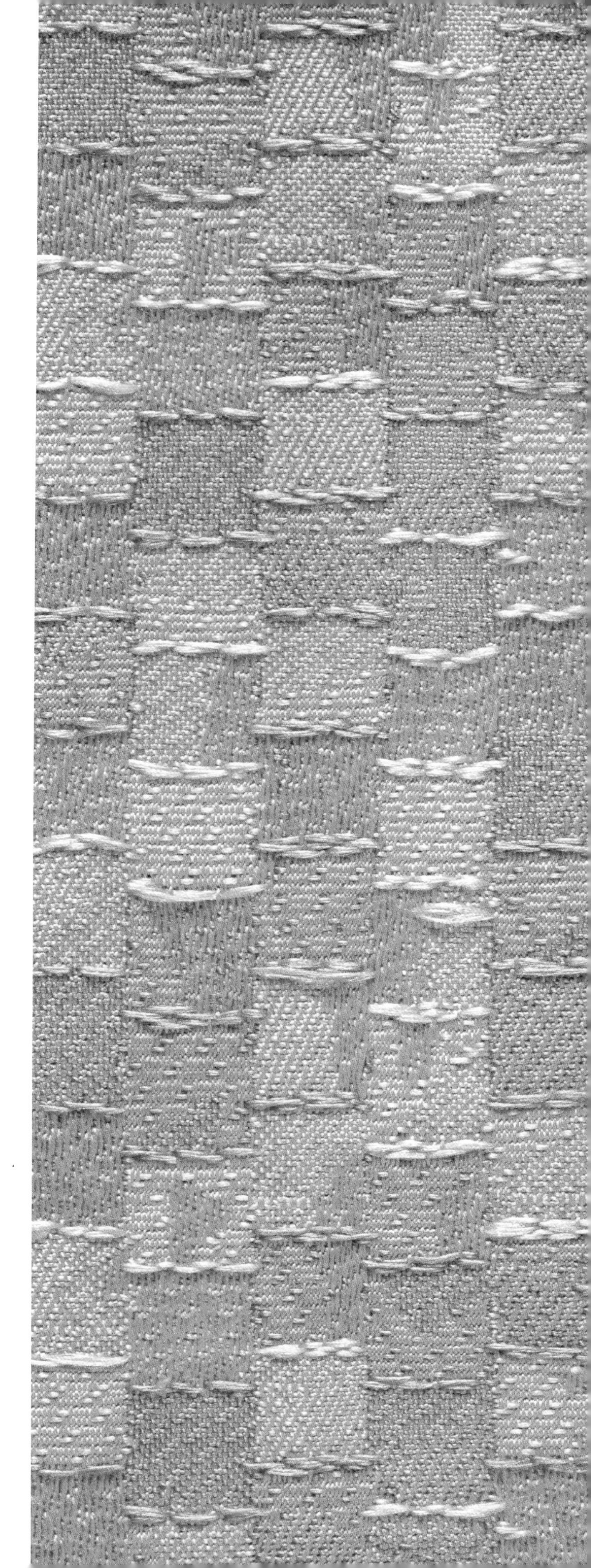

Figure 7.40. Facing page, right: *Artwork and finished sample,* Small Squares, *damassé in silk by AntoinetteStucky, LFS.* This page, above: *weaves drafts;* right: *detail.*

8 Which comes first: technique or design?

Loom and warp come first

Loom and warp specifics define a figured textile. The number of singly controlled ends determines the maximum width of the cloth's design; the number of beams limits the choice of technique. The warp then comes into play: more or less elasticity of materials limits or extends the range of weave repeat sizes in one cloth and the placement of motifs. An elastic warp compensates for smaller/tighter and larger/looser weaves distributed over the design, whereas a nonelastic material such as linen may allow for minimal changes in take-up and weave repeat.

The sett of the warp determines maximum weft float length and weave repeat for ground and figuring effects alike. At a sett of 100 ends per centimeter (254 ends per inch) a 24-end satin generates a 2.5 millimeter float (0.1 inches). At a sett of 10 ends per centimeter (25 ends per inch), the same 24-end satin generates floats 2.5 centimeters (1 inch) in length.

A combination of warp sett and the yarn's cover factor affects the choice of weave structure and family; low cover and sett may make the binding points of satins too visible, whereas a high sett and/or fuller yarn may exclude the use of plain weave. All these elements and more are considered before the designer or artist embarks on a new project.

Design before technique; technique before design

Once the loom and warp have been defined and their potential considered, which step come next: choice of artwork and pointpaper, or technique? Artists often employ figured textiles to create image-based works, whereas those with a background in wovens tend to explore structural and textural opportunities offered by a given technique.

Whether imagery or technique has precedence, each will affect the other. No single technique is an ideal vehicle for every design, and no design can be woven effectively with every figuring technique past or present. Damask is an appropriate choice for artwork with positive/negative areas, shading, and larger motifs. Self-patterning and pattern wefts are excellent for describing fine lines, intricate patterning, and motifs that stand out from a background. Brocading serves for limited areas of discontinuous patterning, especially if fragile or rare materials are employed. Samite and taqueté are potential choices for designs with large areas and no fine patterning that require more color contrast than damask. These two older techniques produce patterning on both face and reverse, whereas their successor in time, lampas, is a one-sided technique and an excellent vehicle for producing most designs, if a loom mounted with two warp series for ground and binding is available (see Chapter 6).

Figure 8.1. Facing page: Pick-up-Sticks II, *plain weave with ground and three pattern wefts and a straight repeat; Helena Loermans, LFS.*

Figure 8.2. Briciola at Night, *lampas with ground and pattern wefts and half-drop repeat; Barbara Shawcroft, LFS.*

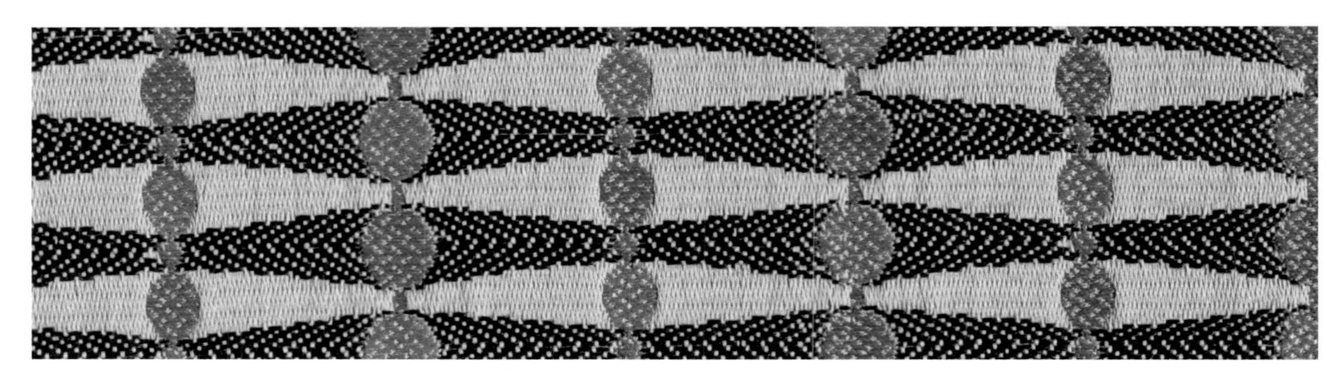

Figure 8.3. Tiefly, *lampas with ground and two pattern wefts and straight repeat; Tuulia Lampinen, LFS.*

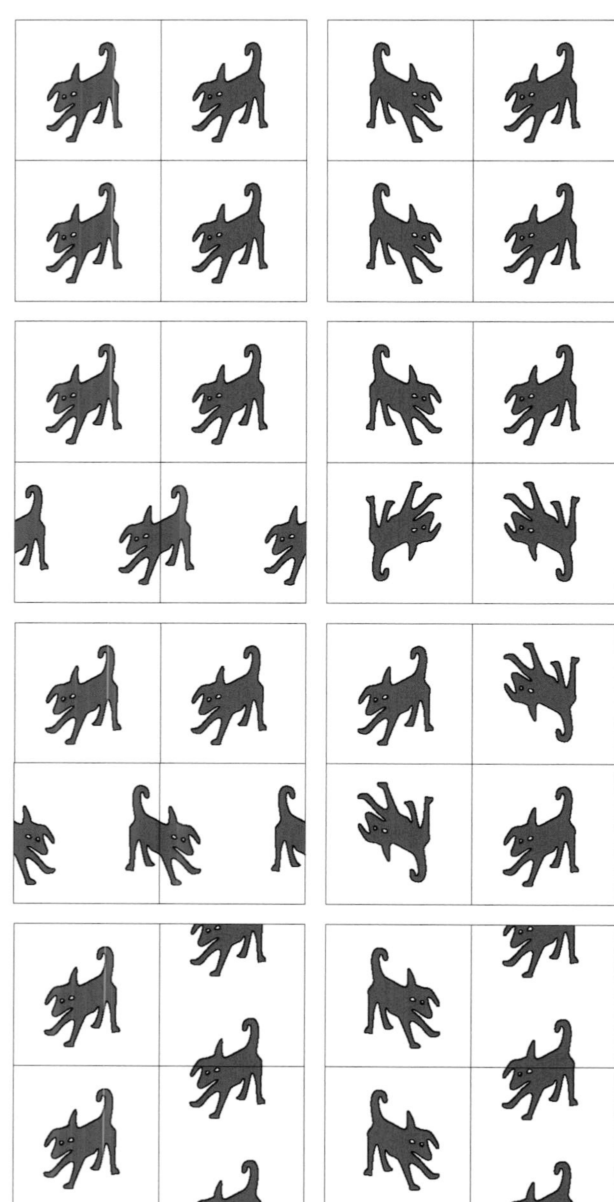

Figure 8.4. *Repeat modes commonly used for decorative figured wovens.*

Figure 8.5. Facing page: Facett, *figured silk with one ground and two pattern wefts and straight repeat; Morgan Bajardi, LFS.*

To repeat, or not to repeat

At one time, the majority of figured textiles was produced on looms set up to weave two or more repeats of the same pattern unit across the width of the cloth (see Chapter 4). Repeating patterns made figured wovens multifocal and decorative by nature, distinct from other two-dimensional mediums such as painting and tapestry.

With the advent of digital technologies, the Jacquard loom has changed, and with it, the identity of figured wovens. Digital looms with single-end control from selvage to selvage permit handweavers and mills to create tapestries, woven paintings, interior textiles, and cloth with non-repeating images. Does this mean repeating patterns will cease to exist? As with so many aspects of our lives, our relation with pattern and how and when we use it is changing because of digital technologies, but if patterning is desired, designer, artist, or mill will be freer to choose scale and number of repeats. Our eyes will continue to seek order, a way to count and measure, to replicate the symmetries and rhythms of nature; we will continue to need the stimulus or security of repeating pattern.

Figure 8.4 illustrates repeat modes typical of traditional figured weaving, starting with the *straight* repeat in the upper left corner. Choice of repeat mode and the distribution of motifs within the repeat's frame can draw the eye to a single element or allow it to move freely between focal points. In the first case, the design is contained by the dimensions of the repeat and has a more emphatic rhythm, in the second the limits of the repeat cease to be apparent and the design "expands."

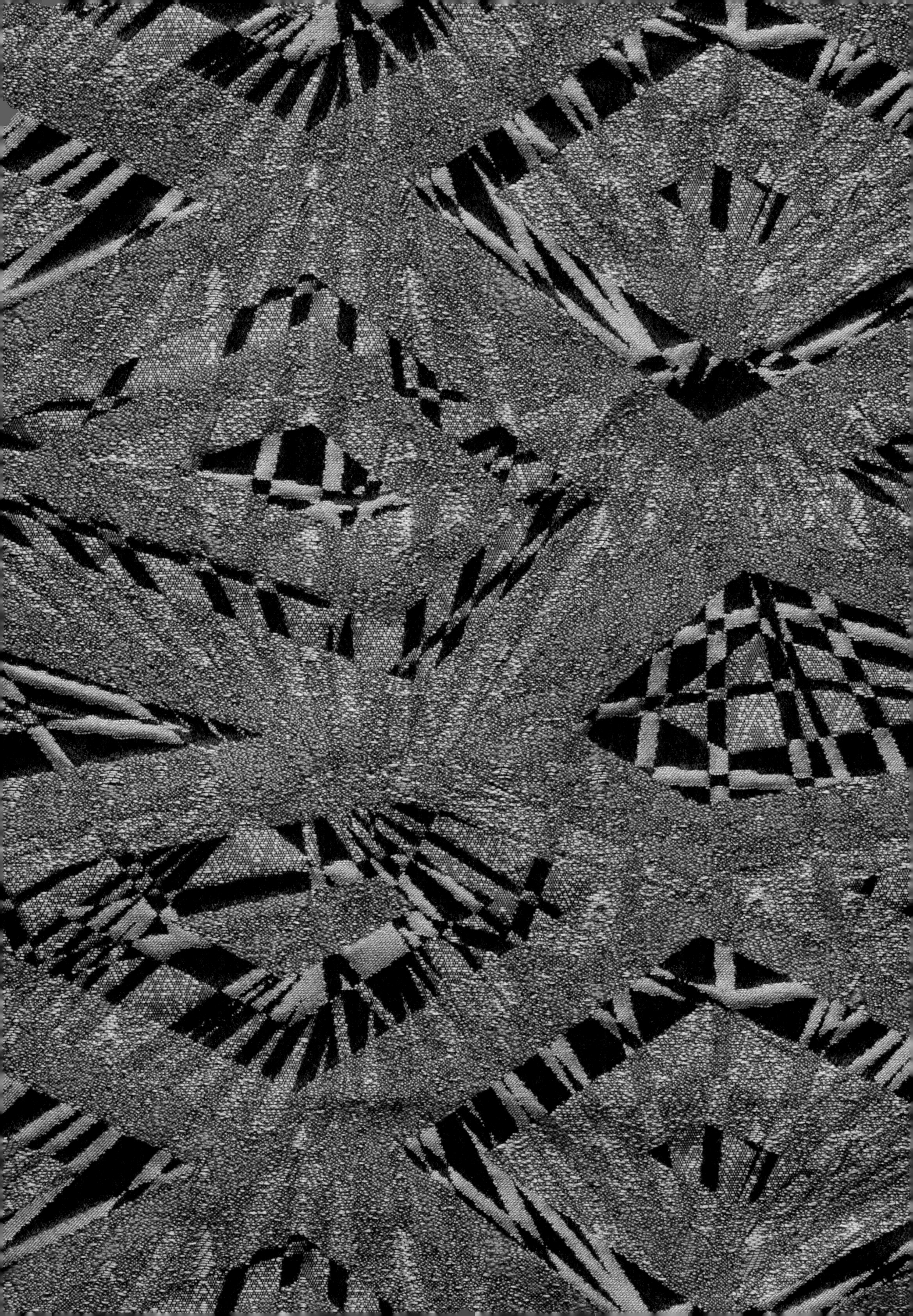

Figure 8.6. Martine, *figured silk with one ground and two pattern wefts and straight repeat; Martine Peters, LFS.*

Trusting the eye

As a project evolves from sketch to pointpaper, simulation, and finally to woven sample, new choices face the designer: Does the contrast between weave areas produce the expected effect, does the distribution of motif and color create the intended flow or interruption of the eye's movement? Are lines, forms, and repeat harmonious or dynamic as these appeared on the pointpaper? If a motif or line appears awkward in a sketch or simulation, it will be more so in the finished project; empty or crowded areas will worsen, and a break in the flow of the repeat will be more emphatic when the textile is woven.

A designer's eye is his or her strongest tool for assessing results. The textile in making can be studied at close range, reduced on the screen, viewed from across the room or through partially closed lids that blur details and bring the overall effect into focus. The eye enables the designer to discern potential in an unexpected result, perceive when a new direction should be abandoned, or confirm a choice.

Facett, shown in Figure 8.5, was "caught" by the designer while she experimented with digital design tools and the original crystal-shaped motifs of her design fragmented into moiré effects and dynamic patterns.

Choosing mistakes

In the course of many projects, the unexpected happens and results differ radically from those projected. Starting afresh may not be an option. Alternative solutions are rapidly tested via simulation or sampling and chosen on the spur of the moment. In the end, the designer's discerning eye can transform an oversight into a fortunate choice.

At times the pointpaper or artwork is inadvertently altered, forcing a designer to abandon the original project and seek a new direction. In *Martine* (Figure 8.6), the encounter of an experienced designer/weaver with a less familiar design tool generated textural twig and leaf patterning; smooth, pebbly, and ribbed weave structures were assigned to the new and accidental effects, and the textile woven.

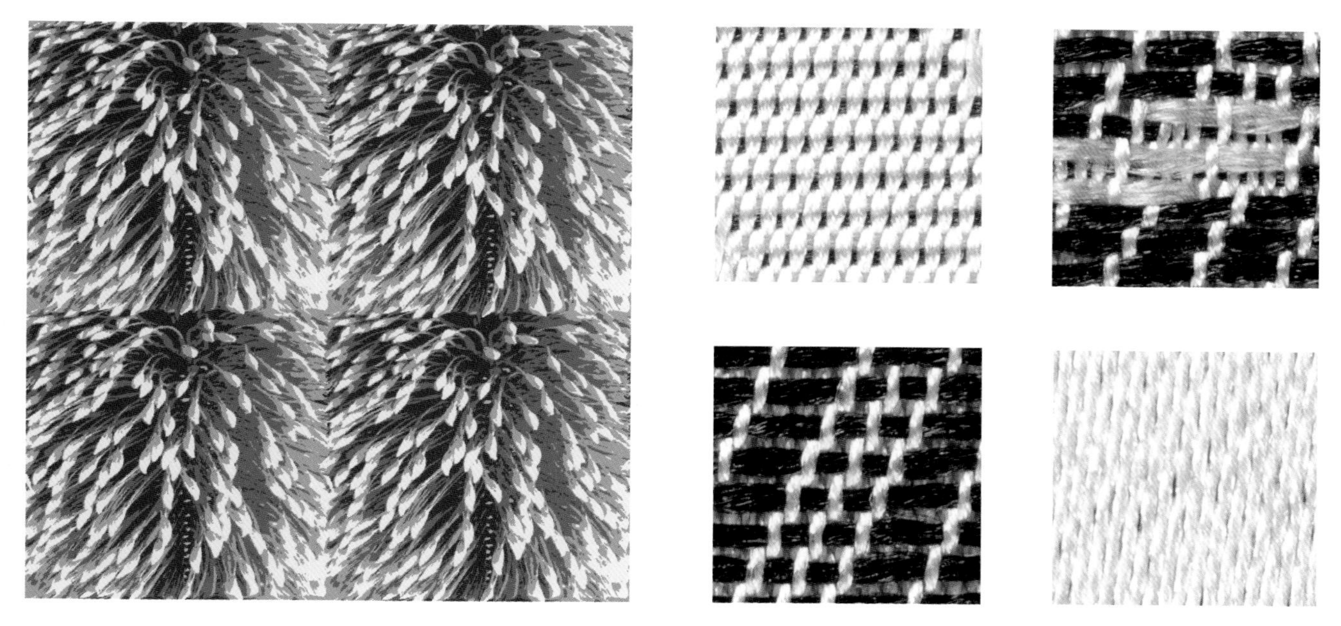

Figure 8.7. Banksia, *damask with self-patterning wefts, Jennifer Robertson, LFS.* Above left: *design with straight repeat;* below left: *woven with a brick repeat.* Above right: *detail of weave effects;* below right: *finished textile at real size.*

The most frequent unforeseen in figured weaving is unequal warp tension (see Chapter 7). Originally *Banksia* was designed with a straight repeat. The vertical axis of the artist's motif, woven with a tight 2/2 basket ground in combination with looser 8-end weaves, generated a problematic degree of unequal warp take-up. By switching from straight to brick repeat, tension was equalized, and the vertical flow of the design was ruptured, adding impact to the finished work.

Threshold of visibility

Digital technology has taught us to think increasingly in terms of resolution: screen resolution and camera and printer resolution. We have come to associate higher resolution with an increase in quality. The sett of fine silk damask is lower than the resolution of the letters on this page, but when used to produce a man's tie and viewed at close range, it will be silky and smooth, and like these letters, details will blur and lose definition when seen from across the room.

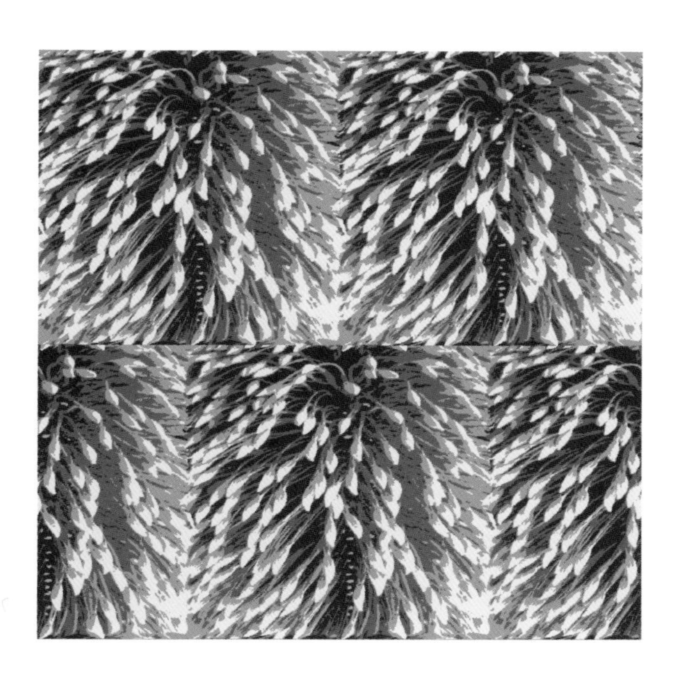

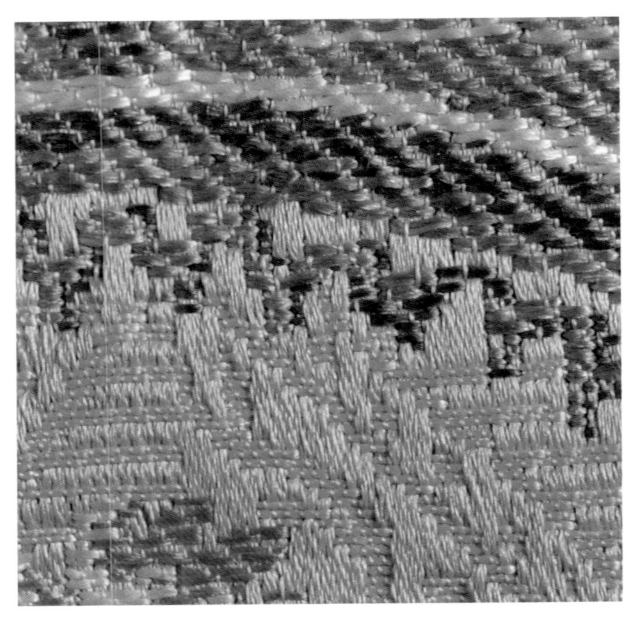

Figure 8.8. *Eighteenth-century silk with figured ground and multicolored brocading. Courtesy of Societé Le Manach.*

When we design a figured textile for a wall, an armchair, or the floor; for slippers, coat, or necktie; the nearest and farthest viewing distance for each use is foreseeable. The distance at which we clearly make out the motifs of a textile is its threshold of visibility.

The motifs of the eighteenth-century yellow brocade are fragmented and hard to read when seen at close range, but smooth out and become legible when viewed from thirty to forty centimeters (twelve to sixteen inches). Seen at a remove of two meters, ground patterning remains visible; at three meters details blur, but color and the sense of movement generated by the motifs' placement and repeat remain.

Ogee is woven on a fine damask warp intended for the production of silk ties. Unlike the historical textile, the pattern is clear at close range; like the older textile, it loses textural definition when viewed from a distance. This difference in threshold of visibility depends on the resolution or sett of the pointpaper, or minimum pattern unit, rather than the ends and picks per centimeter or inch. Textiles with lower setts of pointpaper, ends, and picks will maintain detail and surface interest at a greater distance than either of these silks. Low or high sett does not determine a textile's quality; resolution is relevant to intended use and viewing distance.

Figure 8.9. Ogee, *damask with two self-patterning wefts and straight repeat; Heather Macali, LFS.*

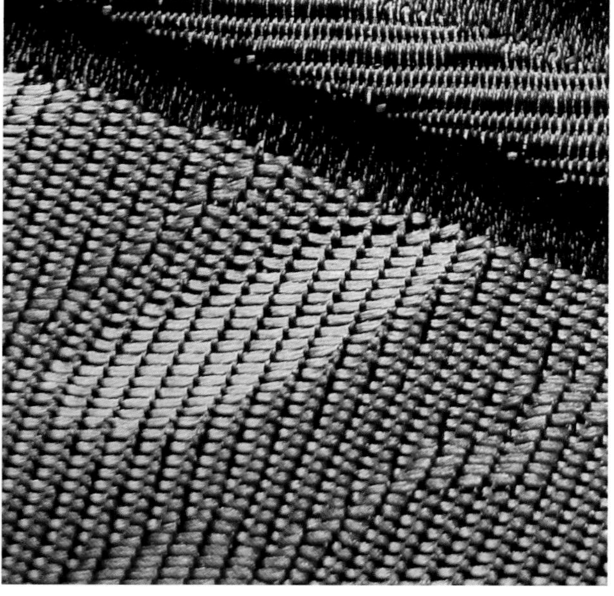

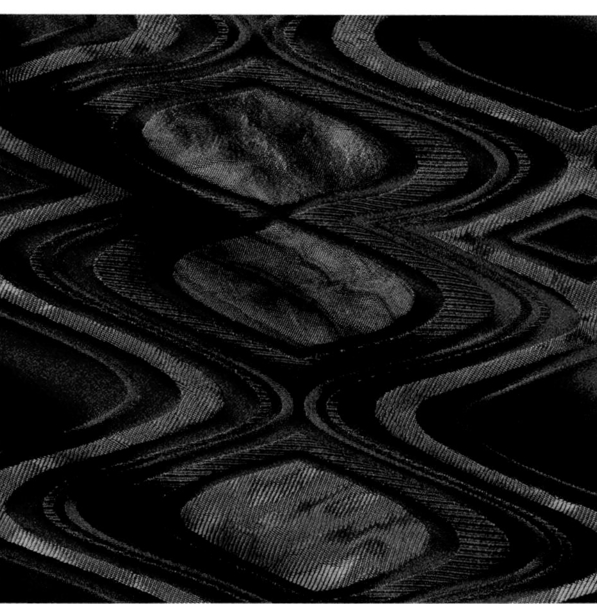

9 Case studies

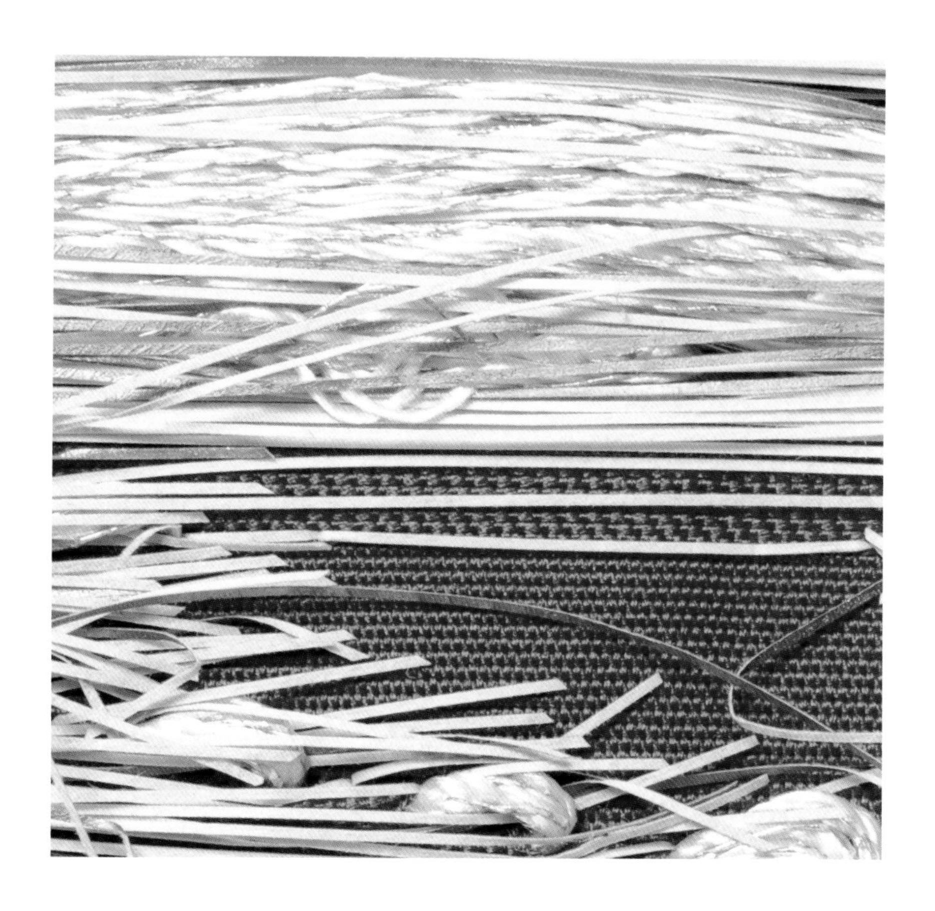

Figure 9.1. Facing page: *eighteenth-century brocaded silk tabby with tassels. Courtesy of Societé Le Manach.*

Figure 9.2. This page: *detail of the reverse of silk, gilt, and lacquered paper figured textile. Courtesy of John Marshall.*

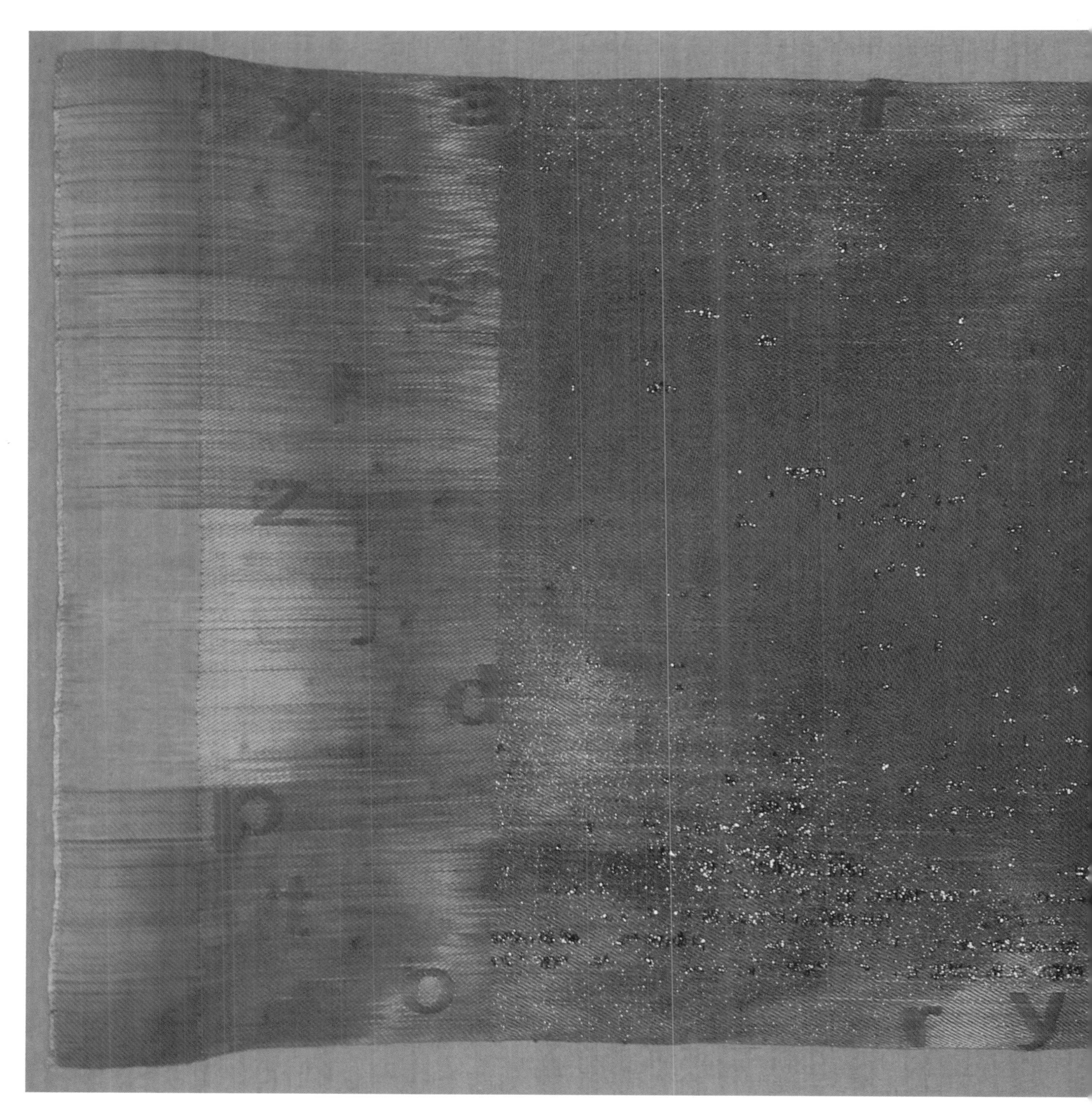

Figure 9.3. Primordial *by Bhakti Ziek. 151cm x 74cm / 59.5" x 29", 2011. Silk, bamboo, cotton, metallic gimp, natural dye extracts. Plain weave and compound satins with two wefts. Courtesy of BZ.*

Primordial and *Genesis*

Both *Primordial* and *Genesis* are based on digital photographs I took of the sky. I almost always start with my camera, then bring the images into Photoshop. I know I wanted *Primordial* to be very soft gradations, and must have started with an overcast sky. *Genesis* on the other hand is clearly a blue sky day. I used two wefts (what I call weft-backed structures) with a satin structure as the basis. I reduce the image to about six colors and then put in the weaves so they shade from warp to weft emphasis. The second weft, a metallic yarn, is a weft-faced, weft-backed structure.

In *Primordial* I used borders of satin against plain weave knowing the plain weave is much tighter and would pucker—I love that effect. Each weaving has an overlay of letters in plain weave. I often use words in my work but I wanted to reduce this to something basic—so letters imply interaction, conversation, communication. At the same time, I have always been fascinated with letters as purely visual forms—the basic mark for me. In *Genesis* the letters are sprinkled here and there (I probably had some format for placement—often I am following a drawn line that is not visible in the finished work) and in *Primordial* I had a full text that I filled into another shape knowing that the final filled form would only have bits and pieces of letters and not be readable. You could almost think of it as people looking at the sky and saying, "do you see a _____blank____ in the sky?" I didn't want it to be identified as something determined.

When I wake up and see blue sky, I know it will be a good day. Also, I realize that no matter where I move, and I have moved a lot, the sky, the same sky, is always above me. I ask myself "where does the sky begin"? We think we can stand and point up but really, it is all around us. In my reductive thinking there is water, earth and sky. I am

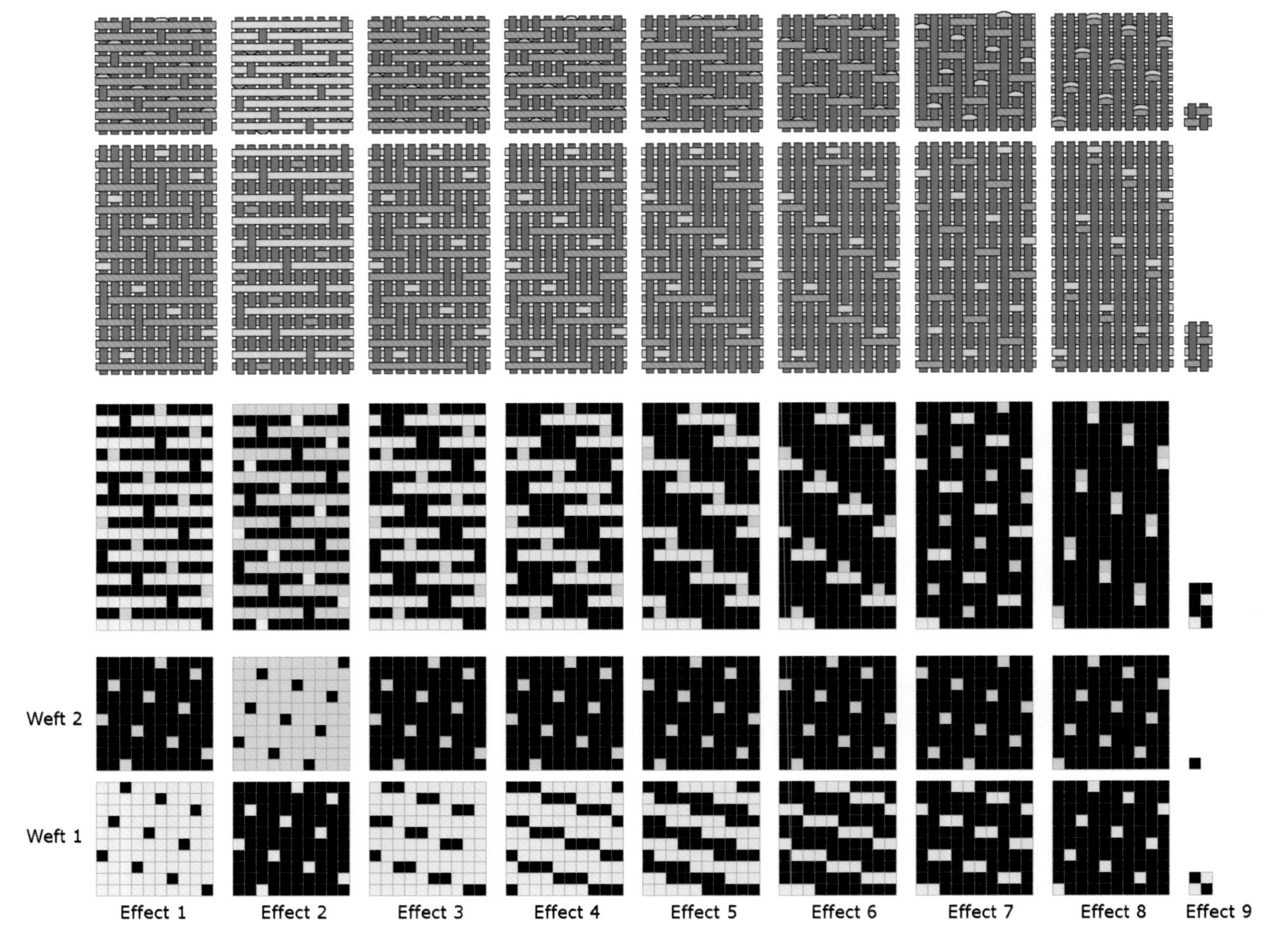

Figure 9.4. *Table showing from left to right the nine weave effects used to weave* Primordial *by Bhakti Ziek. Weave effects 1 through 8 are compound structures composed of one warp and two weft series woven as satins. In effect 9, the first weft series weaves plain weave, while the second weft floats below all warp ends. From top to bottom, row by row: simulations of weaves compacted, simulations of weaves expanded, compound drawdowns, decomposed drawdowns showing the simple weaves used for wefts 1 and 2.*

trying to have a big perspective that reduces things to the elemental.

The warps are dyed with natural dyes. I used madder for *Primordial* and indigo for *Genesis*. I finished the works by mounting them on linen that is stretched over a frame. When I weave I see my cloth under this type of tension—and I love it. The fact that cloth can be a pliable plane is not of interest to me. I am seeing what most people understand as paintings—a firm visual plane. I kept the plain weave beginnings and endings visible to emphasize that it is one element modified throughout—to show transition and transformation. Despite having worked for over forty years as a weaver (is that possible—yes, I took my first weaving class in 1969)—there is still so much to learn and to figure out.

Figure 9.5. Below: *detail with warp and weft aligned as woven and shown in the weave table on the* facing page; right: *detail of* Primordial *as mounted for display with warp aligned horizontally. Courtesy of BZ.*

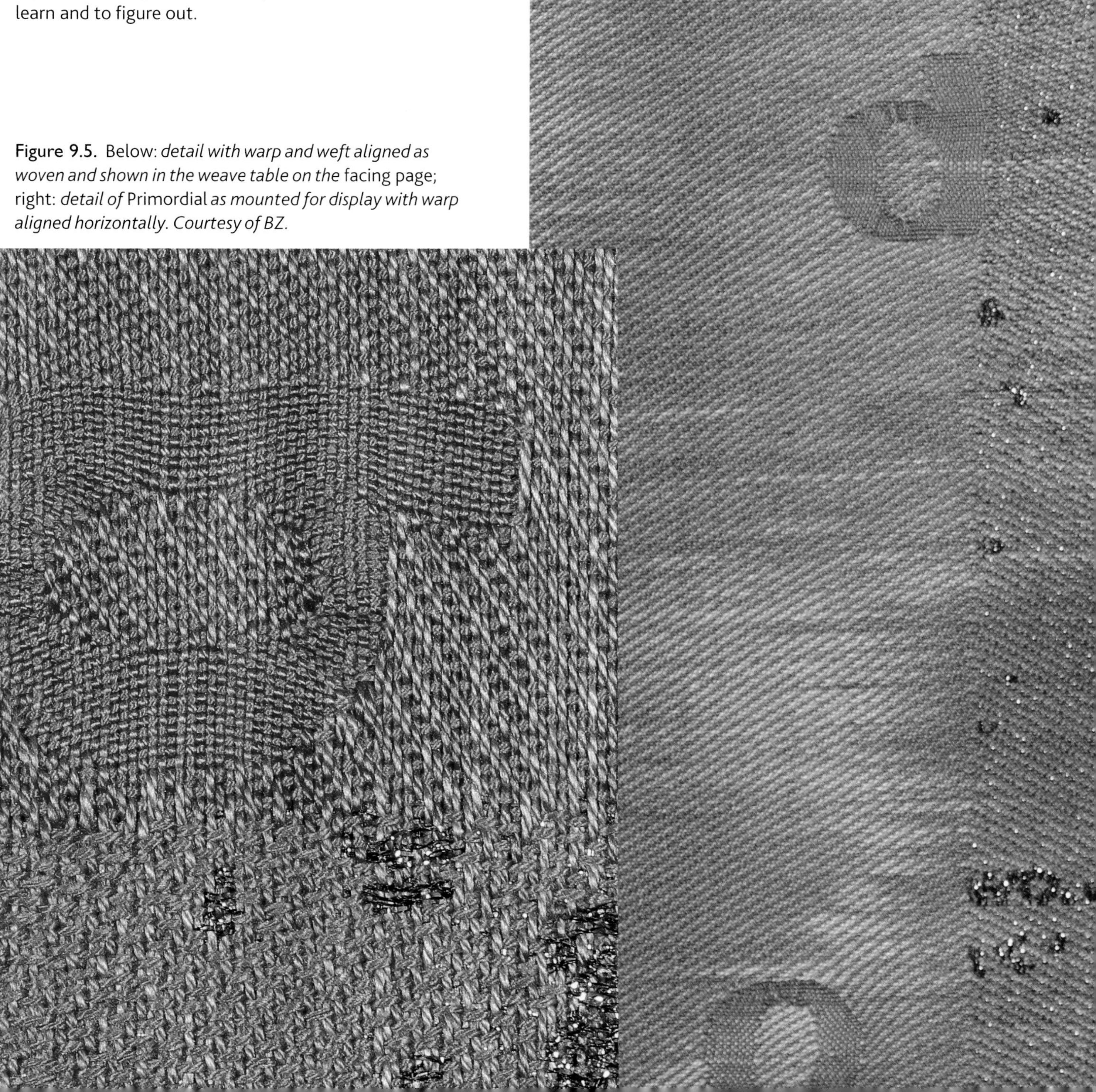

Figure 9.6. Genesis *by Bhakti Ziek, 75cm x 62cm / 29.5" x 24.5", 2011. Silk, cotton, tencel, bamboo, metallic gimp, indigo dyed warp. Plain weave and compound satins with two wefts. Courtesy of BZ.*

In *Genesis* I wove in weft stripes of a different shade of blue just to pop the ground a little bit. Depending on the angle from which one looks at each of these weavings, and the lighting, the changes of structure and weft color are prominent or hardly there. The weavings remind me of a hologram, or one of those Cracker Jack toys that looks different as you move it.

Bhakti Ziek,
artist and educator

Two games, three squares, eight textiles

In the deepest place of my heart I am lazy, so I want to do a lot with as little time-consuming work as possible. I am interested in weaving, in experimentation; my textiles are designed on the loom, not on paper or a computer.

Helena Loermans

For Helena Loermans, weaver, educator, and designer of the eight textiles on the following pages, the Jacquard medium offers an opportunity to experiment with structure, to work out weave puzzles. Loermans employs objects and means that are at hand, be it traditional wooden toys or digital tools, to quickly generate her artwork, which becomes a vehicle for the exchange of woven surfaces. At times experimentation results in a new article for her handwoven production; always it teaches her more about the medium with which she earns her living.

Cinching the pile, the uncut or cut loops that distinguish velvet from other textiles, dictates the construction of the ground weave, whether plain, rib, satin, twill, or patterned. The draft of a typical plain weave velvet in Figure 9.7 shows how the pile is tightly compressed between two ends and two picks. The hand of such velvets is firm, appropriate for upholstery or small accessories.

The artwork for doublecloth, two-pile *Velvet Puzzle* in Figure 9.8, started with two pieces of a child's old-fashioned puzzle; Loermans scanned these and then positioned her motifs on the grid of a digital pointpaper. Next she selected and sprayed the pieces with blue and red, using a virtual airbrush. The loom she would weave on was set up with a ground warp threaded on eight shafts at a sett suitable for plain weave or a tight rib weave construction, and two pile warps series for patterning. Intrigued by the idea of weaving doublecloth velvet, she constructed a ground weave in which all ends of the ground warp and all picks of a first weft series would weave the upper cloth, while the two pile warps wove with a second weft to produce the lower layer. Each pile warp series was controlled by a Jacquard harness to create the two pile effects, drawn in red and blue on the pointpaper.

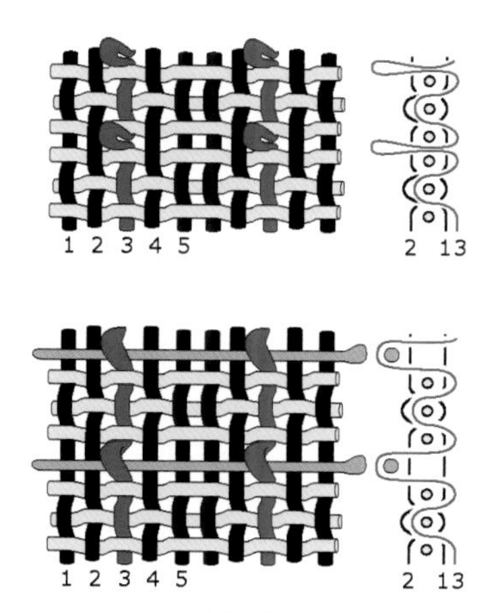

Figure 9.7. *Uncut velvet with plain weave ground.*

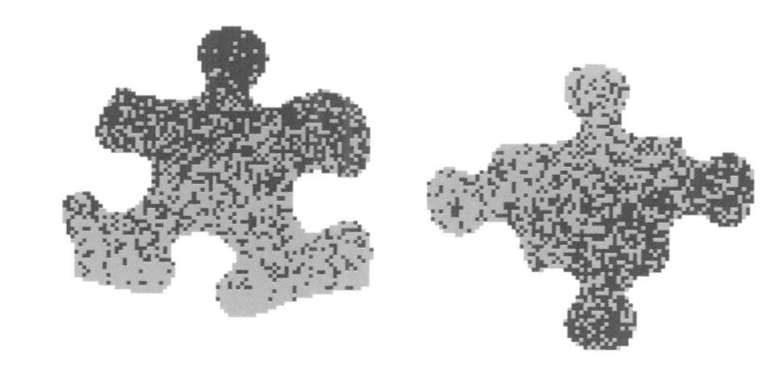

Figure 9.8. *Two puzzle pieces; pointpaper with ground and two pattern effects.*

Figure 9.9. Below: *Drafts for ground and uncut pile effects of a doublecloth two-pile velvet.*

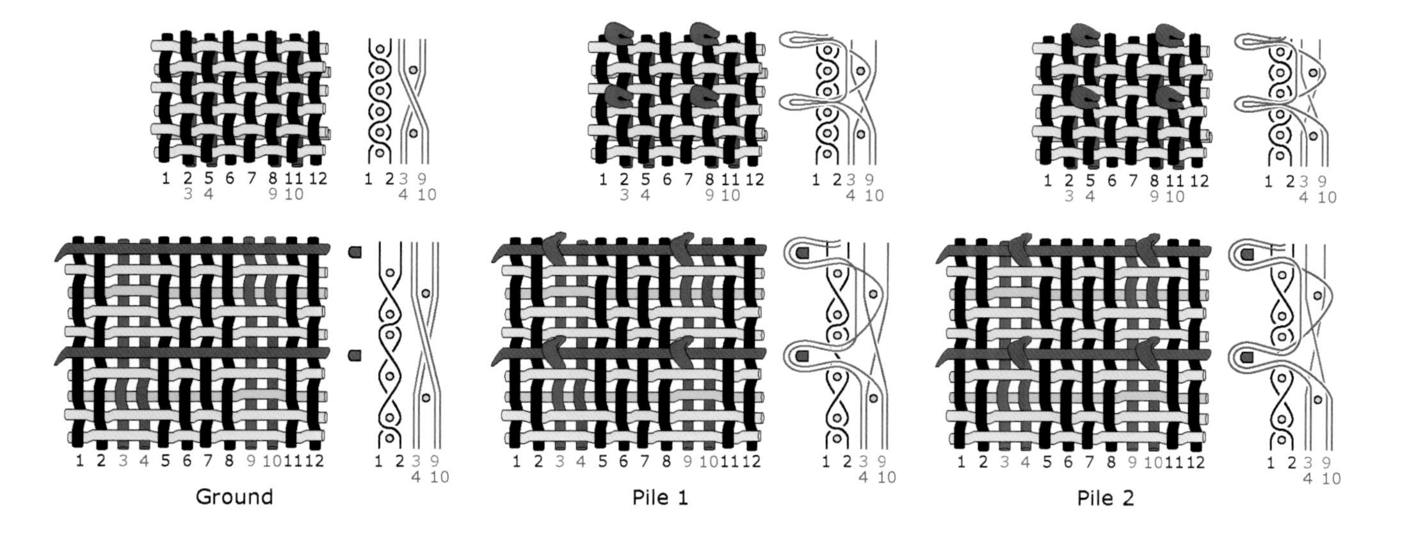

Figure 9.10. Velvet Puzzle *by Helena Loermans, FLS; two-pile doublecloth uncut velvet. Detail.*

The resulting textile is more supple than its plain weave counterpart, but less resistant to wear.

Gradations from warp to weft-faced satin are frequently used to produce shading and a sense of dimension in figured textiles. Loermans employed seven gradations from warp to weft-faced satin 8 to create two geometrically patterned damasks, industrially woven *Damask Squares* and handwoven *Shaded Squares*. In *Damask Squares,* the warp to wefts steps progress vertically, while the color of each row of squares is woven with a single weft color. In *Shaded Squares,* primary colors yellow and red and blue weave singly and then in combination to produce secondary colors orange, purple, and green. The steps from warp to weft satin progress horizontally.

In a third sample, *Clipped Squares,* Loermans inserted discontinuous pattern wefts between the picks of *Shaded Squares.* Colored squares were created by binding two or three patterning picks per weft line above the ground weave, leaving these wefts free to float on the face between motifs. Where a single color was bound to the ground, red, yellow, and blue motifs occurred; when any two weft colors were bound in the same shed, orange, red-violet, and yellow-green squares were produced. Loermans then clipped the floats to create the fringed borders of each square.

Figure 9.11 Above right: Shaded Squares; below: Damask Squares *and* Clipped Squares.

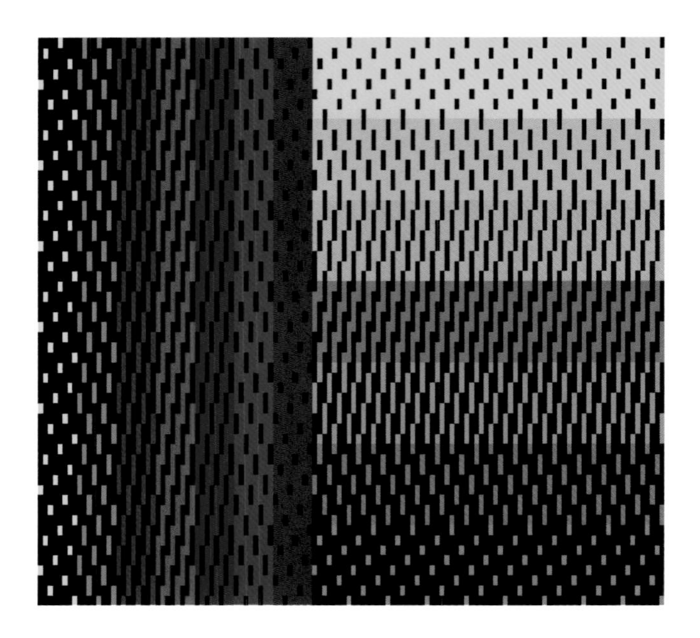

In *I Ching I,* a red warp and a series of white, grey, and dark grey ground wefts were woven in shaded satins to create a geometrically patterned damask based on the pointpaper in Figure 9.12. Twenty-eight figuring picks in white, red, and grey tones were inserted between the wefts of the ground damask to create hexagram-like patterning. During weaving, Loermans selectively skipped or wove the figuring picks to vary the *I Ching* figures.

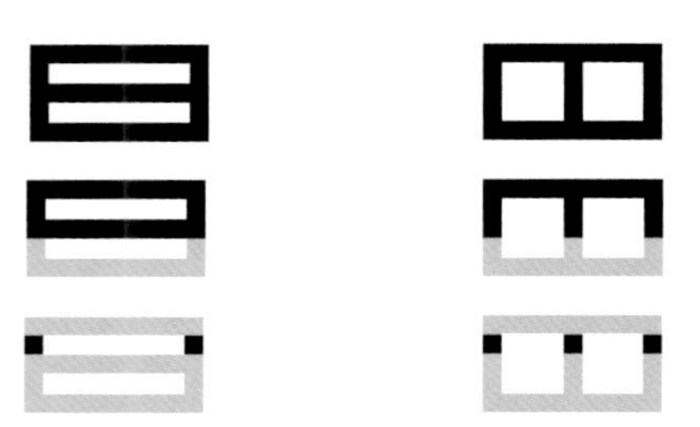

Figure 9.12. Above: *pointpaper for shaded damask ground;* below, left: *detail of woven sample;* upper right: *pointpaper for patterning weft motifs and variations;* below, right: I Ching I, *damask with one pattern weft.*

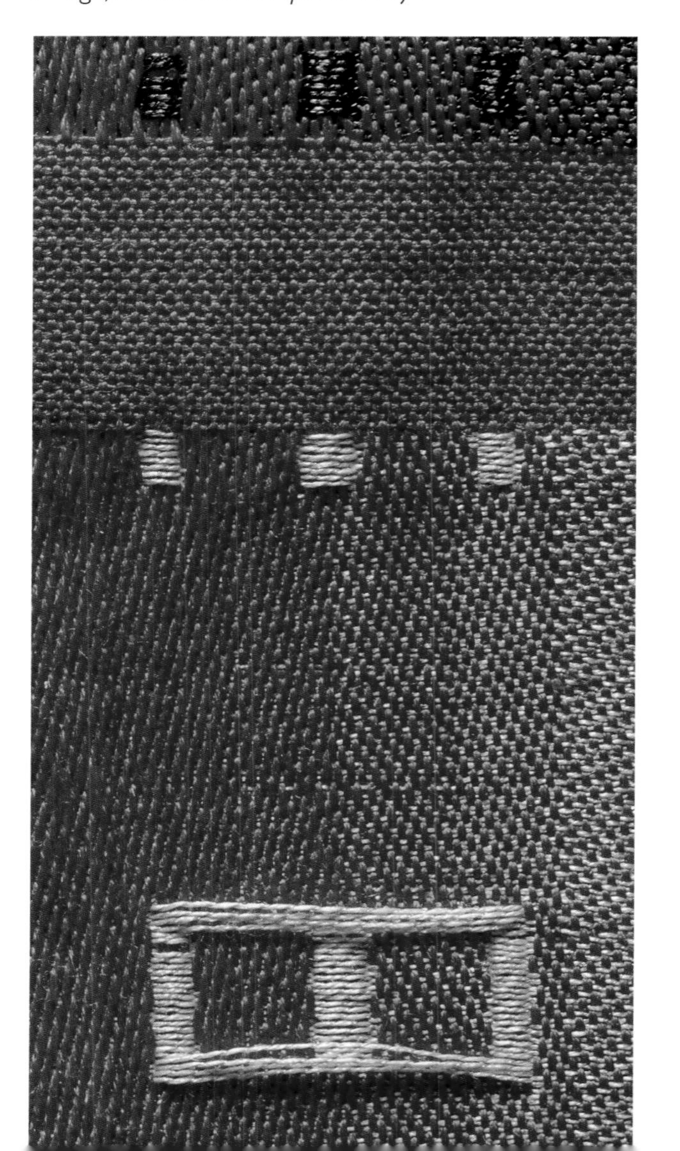

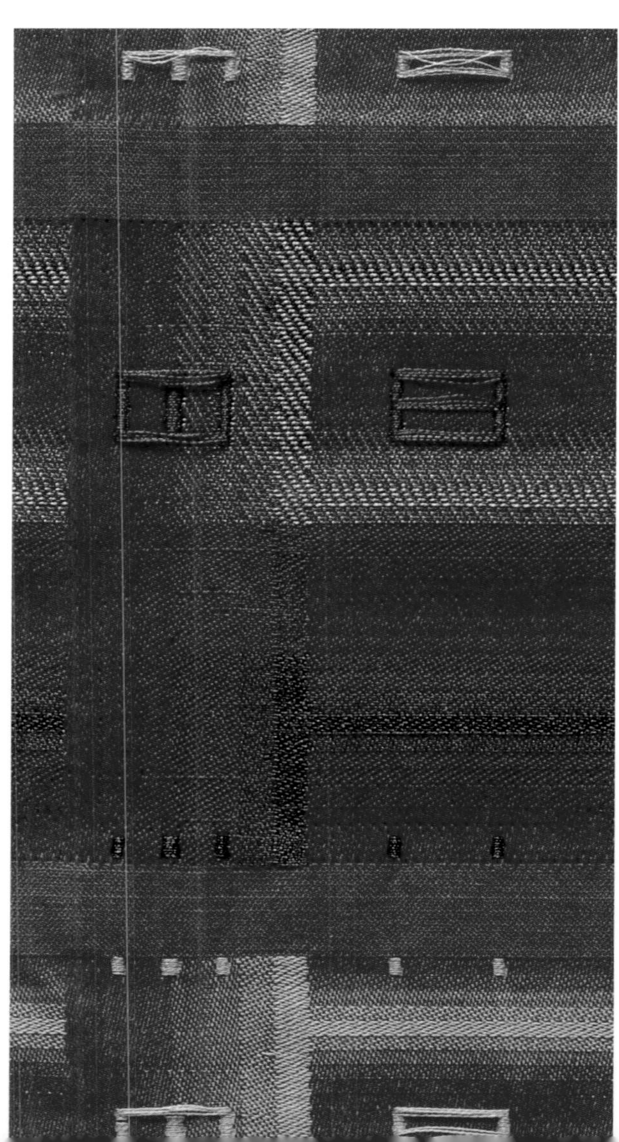

Loermans's experimentation with damask, doublecloth, shaded satins, and patterning wefts was carried one step further in *I Ching II*. Here the same damask ground alternates with horizontal bars of doublecloth, while the selective insertion of supplementary wefts creates hexagram-like patterns above the damask ground and below the upper layer of the doublecloth.

Figure 9.13. Below: *pointpaper for pattern weft insertions; shaded doublecloth damask with pattern weft;* right: *detail,* I Ching II.

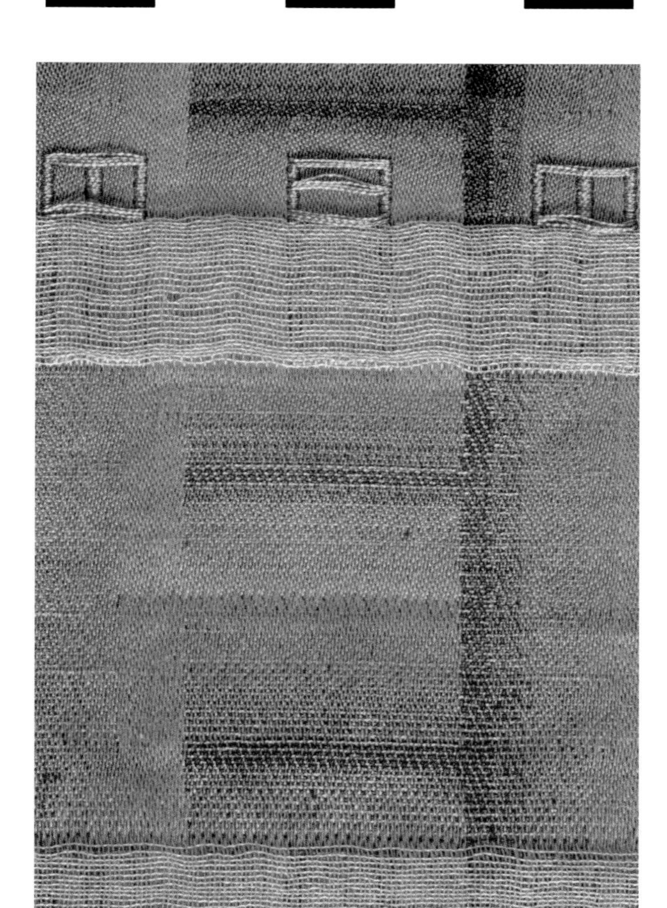

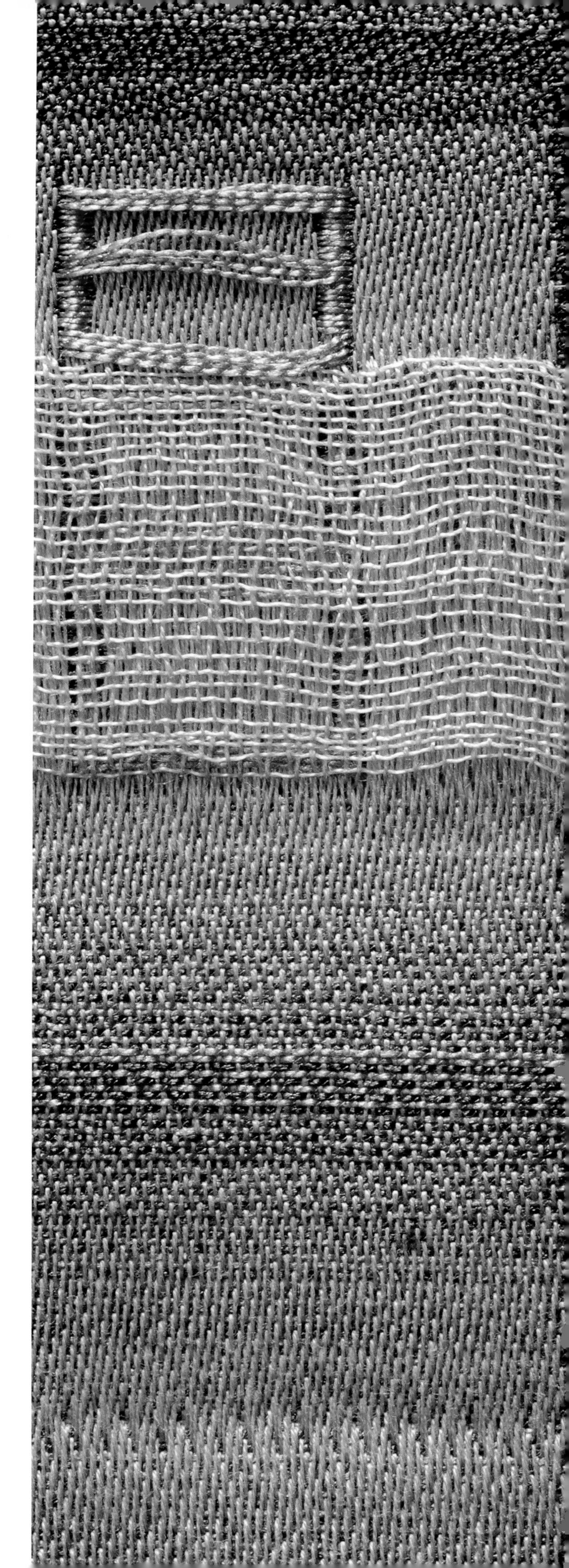

Many weavers prefer the appearance of one binding system to another, while shunning a given weave family. In industrially woven lampas *Pick-up-Sticks I*, a warp satin with one ground weft and three pattern wefts, all three weave families—plain, twill, and satin—are present. The colorful appearance of the textile is produced by use of more than one color within a single weft series, rather than by increasing the number of wefts. In order to economize on pattern weft insertions, the proportion of ground picks to pattern picks is 2 to 1, as can be seen in the weave table in Figure 9.14.

Figure 9.14. Above: *Chair upholstered with* Pick-up-Sticks I, *lampas with one ground and three pattern wefts; photograph courtesy of Paula Oudman.* Right: *pointpaper design;* below: *weaves.*

Effect 1
Effect 2 Effect 3
Effect 4
Effect 5 Effect 6
Effect 7
Warp 1 Warp 2 Warp 1 Warp 2 Warp 1 Warp 2 Warp 1 Warp 2 Warp 1 Warp 2
Pick 5 Weft 4
Pick 4 Weft 3
Pick 2 Weft 2
Pick 1 Pick 3 Weft 1

In a second, handwoven version of the same design, *Pick-up-Sticks II,* a single warp weaves as tabby with a solid-colored ground weft, and is patterned by three weft series bound in a plain weave derivative, twill, and satin. As in *Pick-up-Sticks I,* the colors of the pattern weft series vary in bands along the textile's height to produce a colorful effect, without increasing the number of weft series, which remains constant throughout the textile. The proportion between ground and patterning picks in *Pick-up-Sticks II* is 1 to 1.

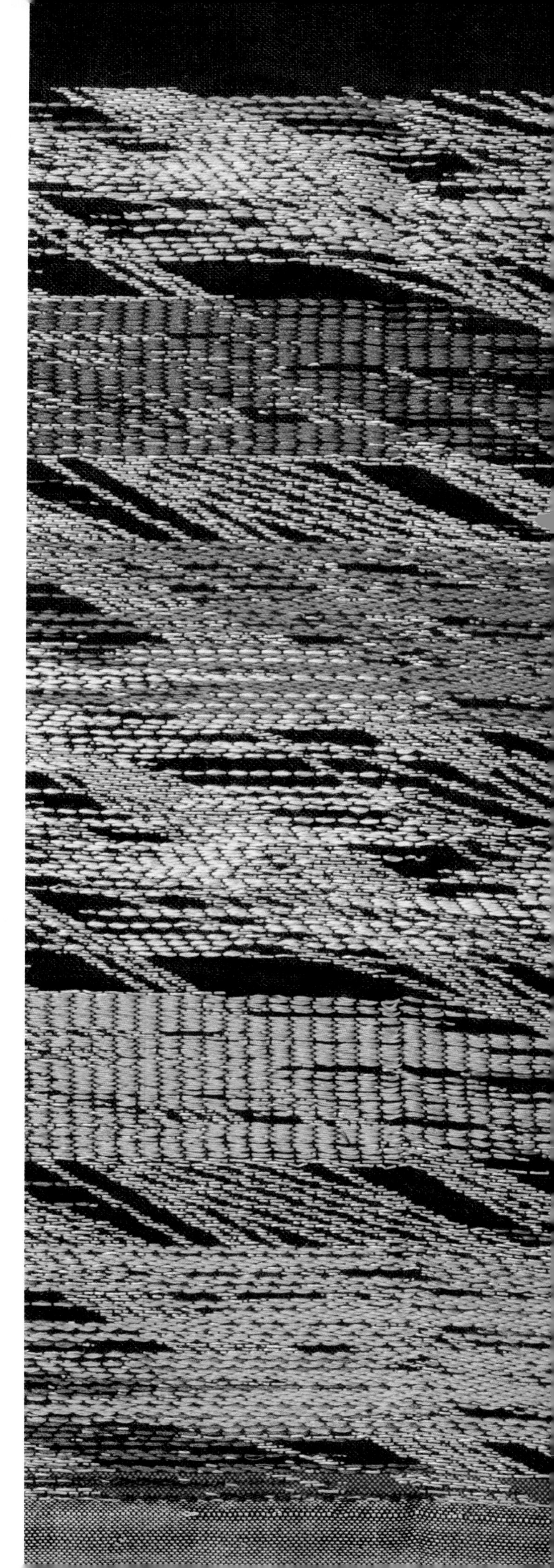

Figure 9.15. *Red/silver colorway; black/gold colorway; detail of black/gold colorway.*

Red July 336: design process, an illustrated narrative

I enthusiastically embrace the iterative capabilities of digital design in the creation of my woven work. The technology allows me to develop and combine networks of patterns derived from traditional weave structures, geometry and my own intuitive sense of order. Motifs built on the circle within the square provide a vocabulary of fluid, lyrical lines and shapes in combination with the grid.

To begin the digital drawing process, templates of different sized circles are designed with mathematical accuracy to acknowledge the edge and emphasize the center of the field. Then the journey of discovery begins as new configurations of shapes and lines emerge from within the regularity of the pattern. Dynamic relationships, orchestrated as segments of templates, are selectively collaged into other fields.

1. Images 01 to 05: this series of images illustrates some of the steps in the evolution of the composition. Circles and ground are selectively combined, outlined, flipped or otherwise manipulated to create interesting interactions.

2. Images 06, 07: these templates are created from weave patterns that have been collaged into a concentric circle field. The precise line and subtle shift of color/value generated by the gradation of twills unifies the field.

3. Image 11: this template contains a variation on advancing twill threading with a random walk treadling and twill tie-up.

4. Images 08, 12a: the graphic quality of the line (composed in the "weave" templates) provides directional movement and contrasting character to the shapes and spaces that they fill.

5. Although it is necessary for me to focus on the visual aspects of my design while working within the virtual realm of the computer, my decisions are predicated on the transformation of pixel to thread. Using thin linen in the warp, two wefts were chosen to create a textural contrast of shine and weight. The sett of the warp was much

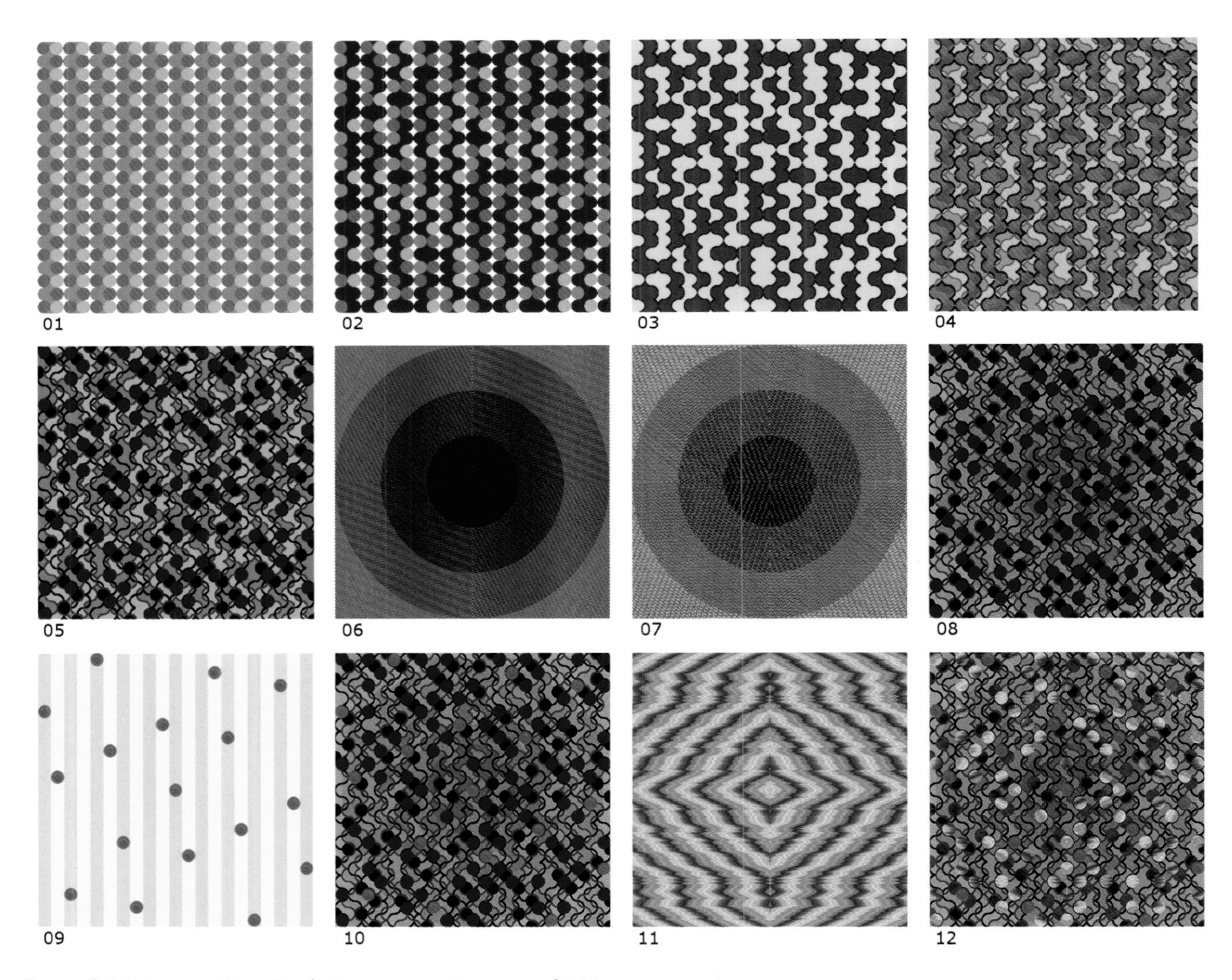

Figure 9.16. *Images 01 to 12: design process. Courtesy of JLM.*

denser than the weft and in this weaving I introduced a third weft, a thick nylon cord, to accent selected circles. Image 09: because the cord would affect the warp take-up differently than the two continuous ground wefts, the alignment of the motifs had to be carefully considered, as illustrated in this template.

6. When the composition is finished, I concentrate on color and the impact of color blending resulting from the intersection of warp and weft during the process of weaving. Templates of warp painting and striping in combination with different wefts allow me to explore many permutations before arriving at my final color simulation: image 13.

The finished weaving, a complex interplay of abstract systems is intended to provide a rich visual experience that invites contemplation and poetic interpretation.

Janice Lessman-Moss, educator and artist

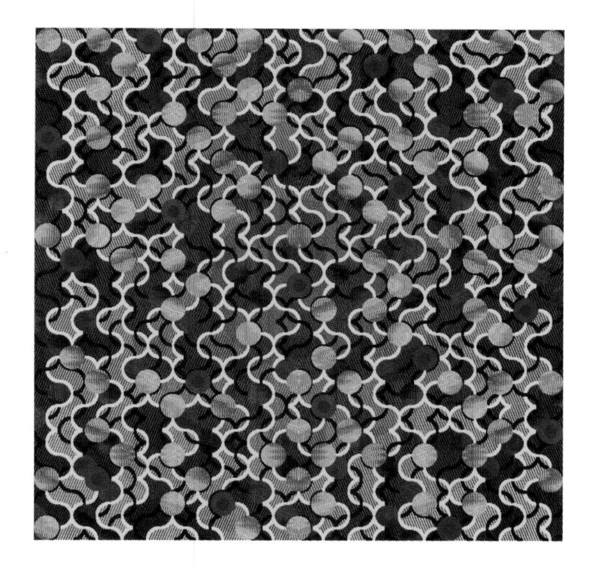

Figure 9.17. *Image 13: simulation of* Red July. *Courtesy of JLM.*

Figure 9.18. Red July: 336 *by Janice Lessman Moss, © 7/04. 140 cm x 142 cm / 55" x 56". Woven on a TC-1 loom with eight modules.*

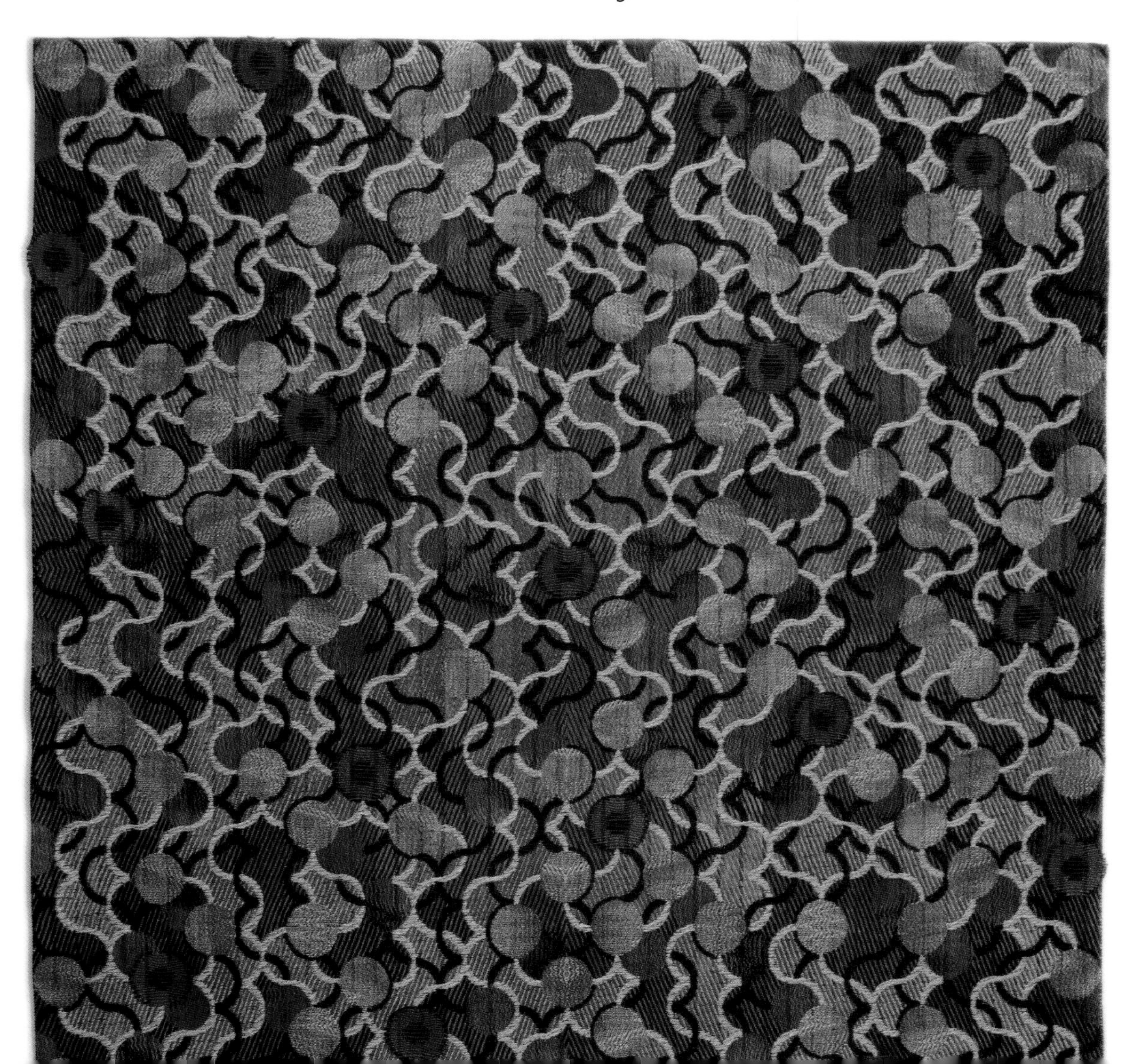

	Effect 1	Effect 2	Effect 3	Effect 4	Effect 5	Effect 6	Effect 7	Effect 8
Weft 3	All up	All up	All up	All up	All up	All up	Plain weave	All down
Weft 2	Warp satin 64 shift 23	Weave field 07	Ogee weave field	Warp satin 64 shift 23	Warp satin 64 shift 23	Weft satin 32 shift 23	Warp satin 64 shift 23	Ogee weave field
Weft 1	Weave field 06	Warp satin 64 shift 23	Warp satin 64 shift 23	Ogee weave field	Weft satin 32 shift 23	Warp satin 64 shift 23	Warp satin 64 shift 23	Warp satin 64 shift 23

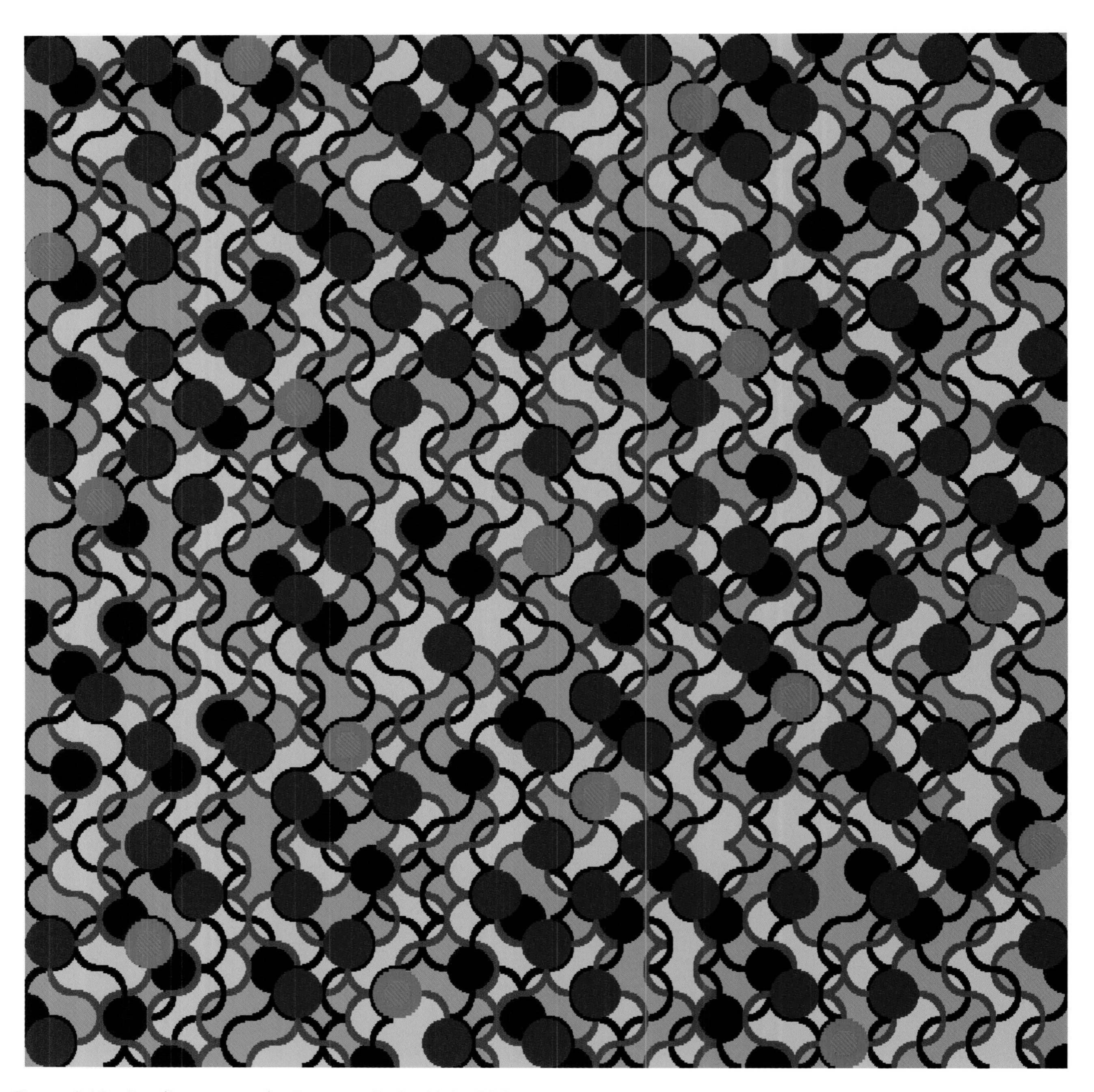

Figure 9.19. *Reading note and pointpaper for* Red July: 336.

In the detail of the finished work, each of the eight weave effects chosen by the Lessman-Moss can be discerned. The painted linen warp is sett at 78 epcm/30 epi; a first weft series in light green spun rayon alternates pick-pick with a second weft in navy cotton, for a total of 78 ppcm/30 ppi. Alternately, the two complementary wefts appear on the surface in effects 1, 2, 3, and 4 to create the principal structures of the work using the weave fields in Figure 9.16. In weave effects 5 and 6, wefts one and two respectively are bound in weft satin 32 to create the raised lines of the patterning. A third weft, a dyed nylon cord, discontinuous in the height and width of the work, weaves circular motifs in tabby, with weft floats at the center. The cut tips of the pattern weft series are sealed with acrylic paint.

Figure 9.20. *Details of* Red July: 336. *Courtesy of JLM.*

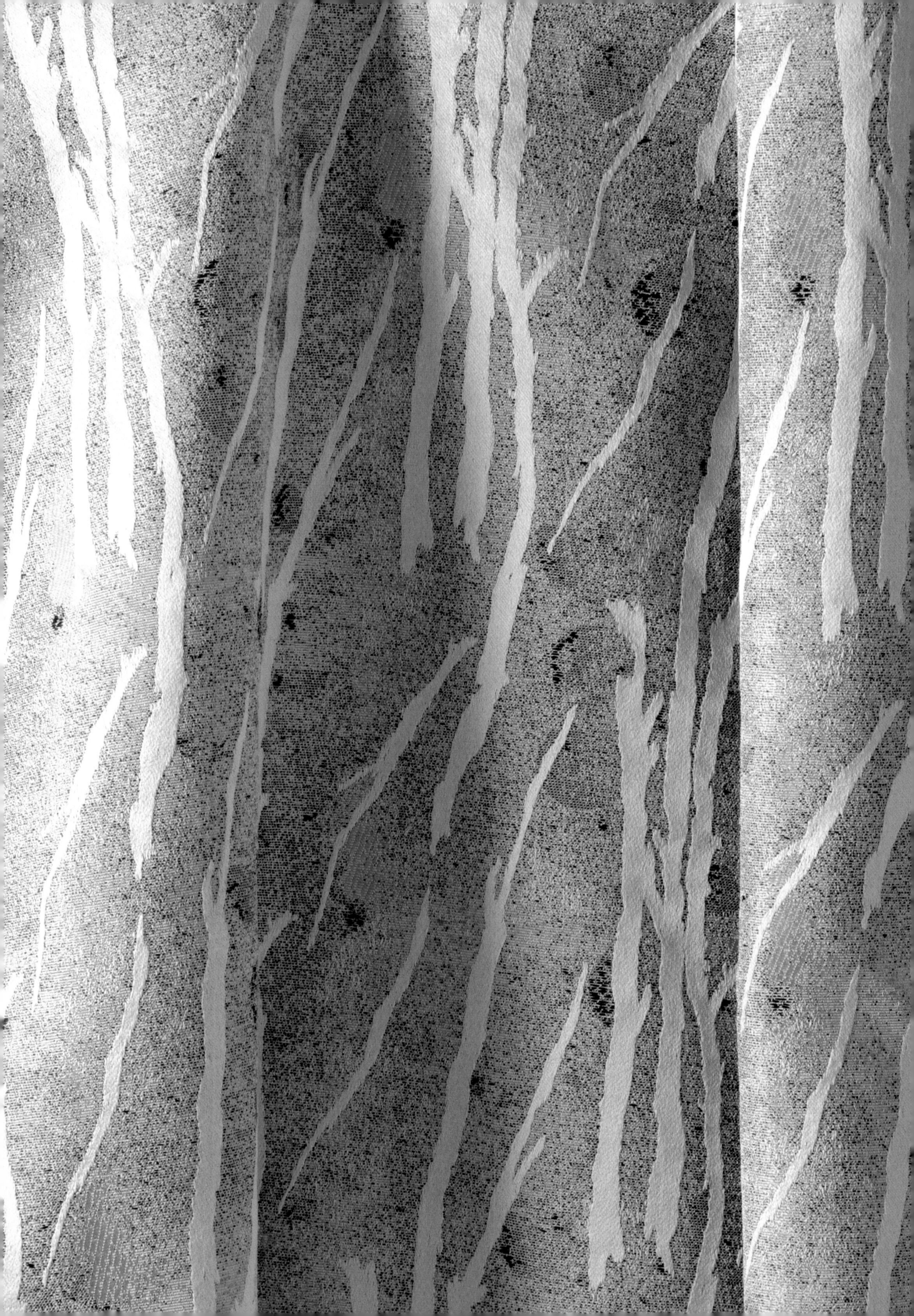

Fast design, fast weaving

The previous works, and many that follow, are the fruit of artists' research and manual execution. The artists were relatively free from constrictions that limit design and weaving time, as well as use of materials, in industrial textile production.

Trees is an industrially woven interiors textile that in all ways respects criteria of fast design, fast weaving, and minimum use of materials (albeit of very high quality). The digitally generated artwork is the result of quickly elaborated photographic images found on the Internet. The background of the design was created by reducing the total number of colors to three of four shades and then manipulating these with a digital airbrush. A weave was assigned to each color, bringing the warp or one of two weft series to the surface, but ultimately, it is the exchange of pixels that determines both appearance and structure of the ground. Intuitively, the designer positioned a motif of leafless tree trunks on the ground; the overall effect reflects the spontaneous design process. The five weaves assigned to the five colors of the pointpaper are classic choices for damask with self-patterning wefts: warp and weft satin 8 and 2/2 basket are simple weaves, whereas the alternating yellow and black weft series are the dominant elements in two compound weaves. In effect 4, the yellow is bound in a low 16-end weft twill, while in effect 5 the black weft is bound in a 32-end weft satin. Both of these self-patterning wefts float above a tight 1/2 rib bed.

The design process and weave solutions, so different from those adopted for the work that follows, produce a textured ground similar to that of the industrially woven painting *Let Go*.

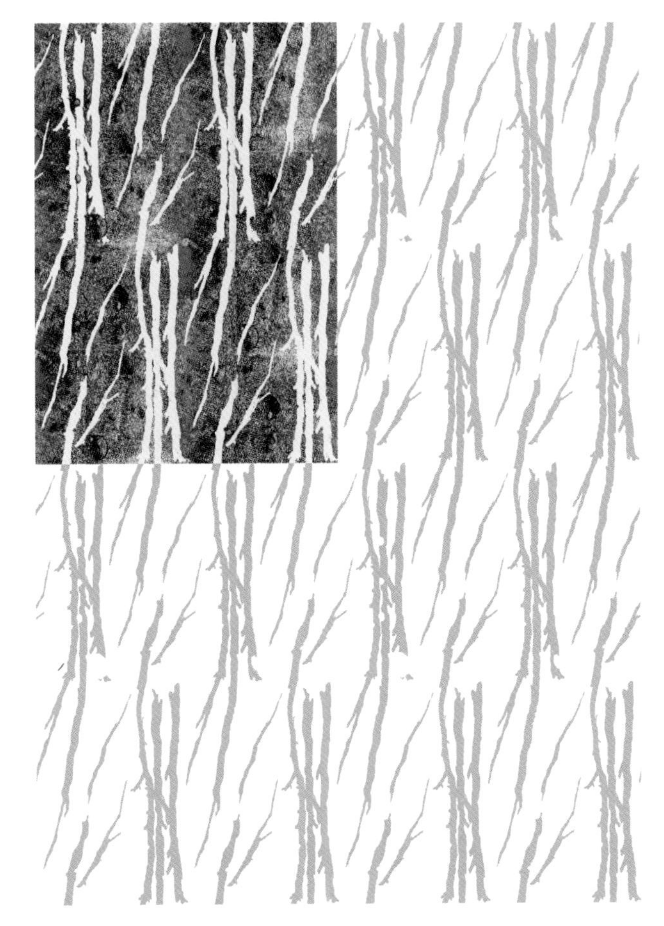

Effect 1 Effect 2 Effect 3 Effect 4

Effect 5

Figure 9.21. Facing page: *figured silk textile woven on fine industrial damask warp at 112 epcm (285 epi) with 32 picks per cm (81 ppi),* Trees *by Matthias Holz, LFS.* This page, clockwise from upper right: *pointpaper and repeat; detail; weaves.*

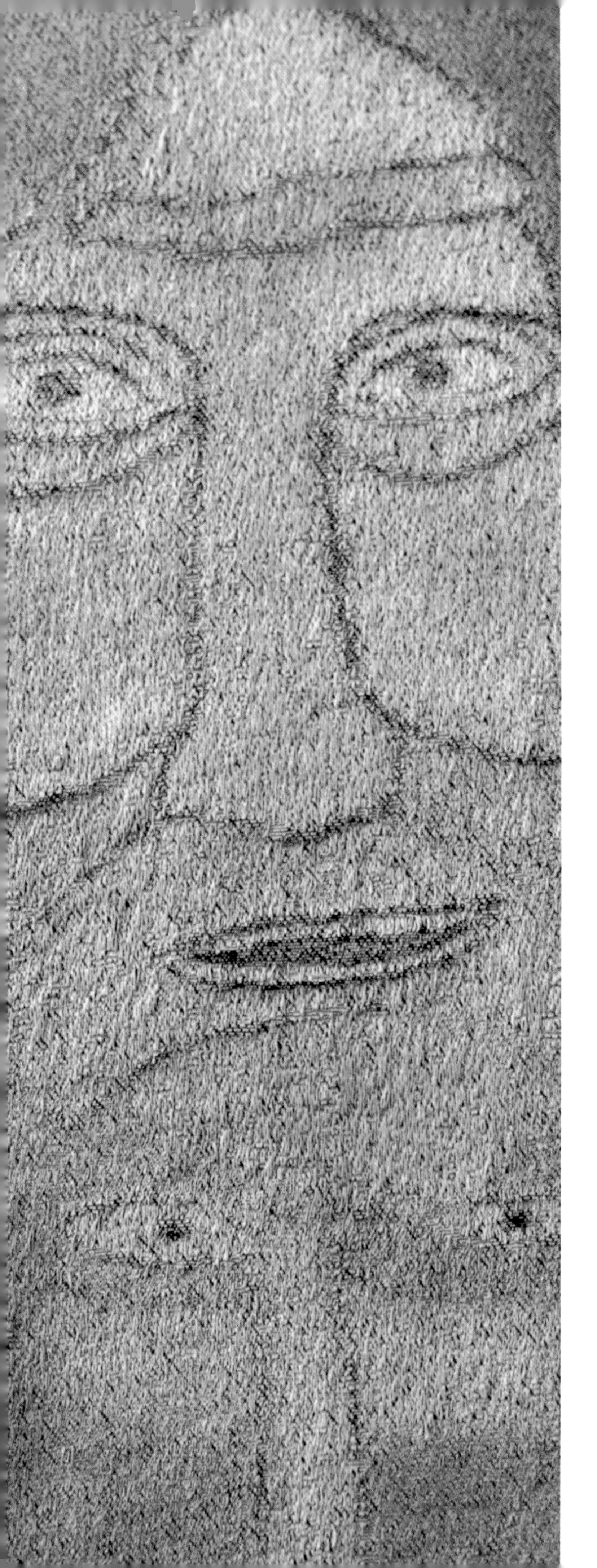

Watercolor, woven canvas, finished painting

Let Go had both a different beginning and end than the other wovens examined in this chapter. Weave patterning in this work constitutes one step in a process that produced a textile in which the exchange of weaves is neither the distinguishing factor nor the final objective of the maker's endeavors. Bethanne Knudson's creative odyssey started with a mental image, a softly gradated watercolor, and ended with the transformation of a weave-patterned "canvas" into a woven painting.

Art creates its own reality. Few things in life are as satisfying as time spent in that reality.

The image-based work I produce consists of capturing, like a freeze frame, a single imaginary moment from a narrative. I love the theatrical, the surreal and the ephemeral. I am interested in the conscious, the sub-conscious and the unconscious. I hope the objects I make will engage the viewer. It is the viewer, ultimately, who completes the work.

The image of *Let Go* is built from a woven structure. The figure emerging from the woven structure is magical; producing it, methodical. I find this aspect of weaving especially satisfying.

The weaves create *imprimatura* or underpainting, where the combination of warp and weft produce a gray scale drawing, in which the weaves build detail. Once off the loom, I work back into the piece, building layers of color using direct application of fiber-reactive dye. The colors have shifts only in hue—all the value shifts are from the woven ground.

Bethanne Knudson, April 2012

Figure 9.22. *Detail of* Let Go *by Bethanne Knudson, painting on weave-patterned canvas. Courtesy of BK.*

Figure 9.23. *Initial artwork, a watercolor, for* Let Go. *Courtesy of BK.*

Using digital technology, Knudson transformed the watercolor in Figure 9.23 into a black to white tonal rendering of the painted image. She then reduced the digital image to seven shades, each of which would be associated to a weave structure.

From the range of warps available at the Oriole Mill, the artist chose an unbleached, mercerized, long staple giza cotton set at 77 ends per cm/195 ends per inch to be woven at 20 picks per cm/50 picks per inch with a black weft, materials that would allow her to produce the *imprimatura* that contained a trace of the watercolor's imagery. A relatively fine, uniform surface was required to facilitate an even absorption of the dye she would apply with a brush, while the variations from light to dark generated by combinations of weave, warp, and weft would either deepen or lighten the colors produced by the fiber-reactive dyes she would use.

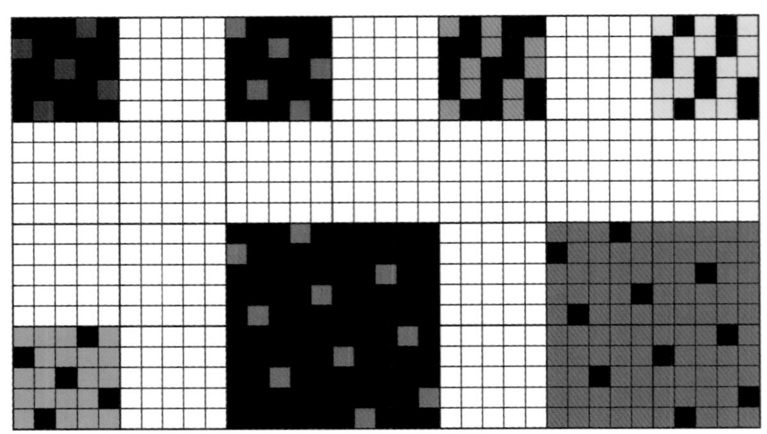

Figure 9.24. Left: *The seven satins used to weave* Let Go; below: *weave-patterned ground used as an "underpainted canvas" for the final work. Courtesy of BK.*

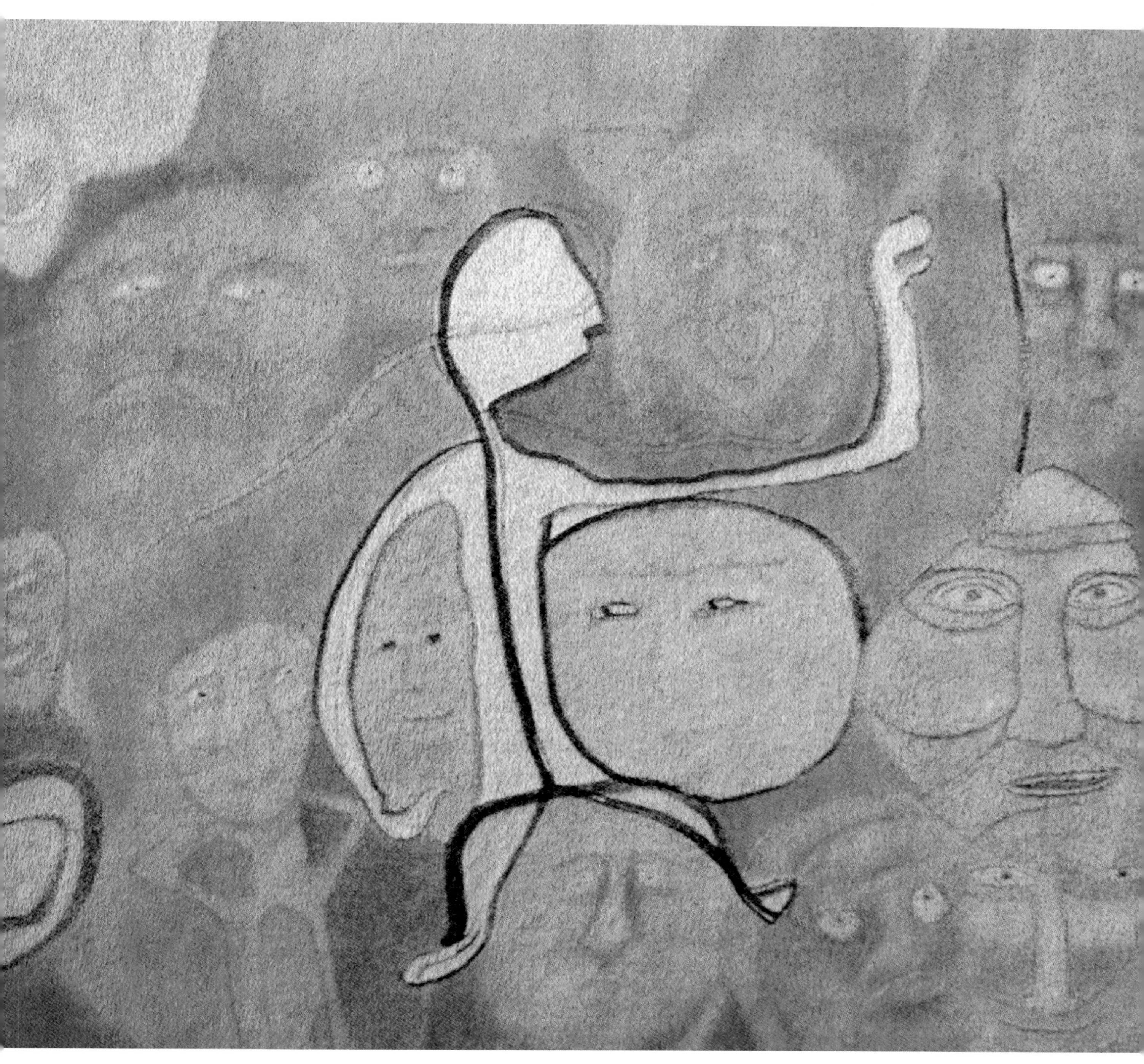

Figure 9.25. *Weave-patterned painting* Let Go, *2008, by Bethanne Knudson. 58.5 cm x 56 cm / 23" x 22", cotton dyed with fiber-reactive colors. Courtesy of BK.*

Satins, which present the smoothest surface of all weave families, were the ideal choice for creating a damask-like palette of seven simple weave effects, ranging from light to dark. Using a progression of warp to weft-faced 5-end satins, Knudson drafted five compact structures; to these she added warp and weft satins 10 that would correspond to the lightest and darkest areas of the painting. Next, the seven structures were associated to the seven shades of her pointpaper.

With weaves and the tonal range these would produce in mind, she defined gradual transitions from warp to weft-faced surfaces in some areas, while augmenting the warp-to-weft, light-to-dark contrasts in others. Adjustments made to the pointpaper, she then wove a weave-patterned ground for her painting.

The last stage of her process was the actual painting of the canvas. Following the shaded weaves and referring to the initial watercolor, Knudson brushed fiber-reactive dyes onto the woven surface. As can be seen in the photograph of the unpainted canvas, the predominantly warp-faced cloth allowed her to use subtle colors that would be visible on the light-colored warp.

Figure 9.26. Facing page, from left to right: *two cotton damasks,* Frost and Skate Marks, *by Robin Muller.* Above: *photograph of Warming Hut. Courtesy of RM.*

Site- and event-specific textiles for the winter season

The two damasks on these pages were designed by Robin Muller, professor at NSCAD Division of Craft and Design, to serve as upholstery material for a warming hut at the 2011 Canada Winter Games. *Warming Hut,* an interactive environment, shelter, and social space, was a public art project created by Architextile Lab as part of a collaborative research project on electronic textiles in architecture. The work group included artists, architects, engineers, and students from NSCAD and Dalhousie Universities.

The cone-shaped structure was made of eight aluminum elements that fit together like sections of an orange. The exterior was covered in wood, clear plastic, and brightly colored PVC-coated fabric. The interior was made warmer by the textiles covering walls and heated seating, while a snowflake chandelier changed color in response to visitors in the hut. When visitors sitting on the central bench of the hut placed their hands inside warming gloves, a hidden heartbeat monitor caught and amplified the sound of their hearts and triggered softly pulsating light.

As often happens with temporary structures that prove useful and attractive, the hut has been reinstalled for a second season and will warm skaters and observers for an indefinite number of winters to come.

Both textiles were woven in cotton on a 40 ends per cm/102 ends per inch white warp, at 18 picks per cm/45 picks per inch, in alternating colors of blue and green. Photographs of two forms of ice—natural formations of frost flowers and skate marks generated by human activity—were sources for the artwork. The frost flower image was blown up to a large scale; colors were reduced to four tones from white to blue, each of which represented one of a series of 5-end shaded satins shown in the pointpaper design and relative reading note in Figure 9.27.

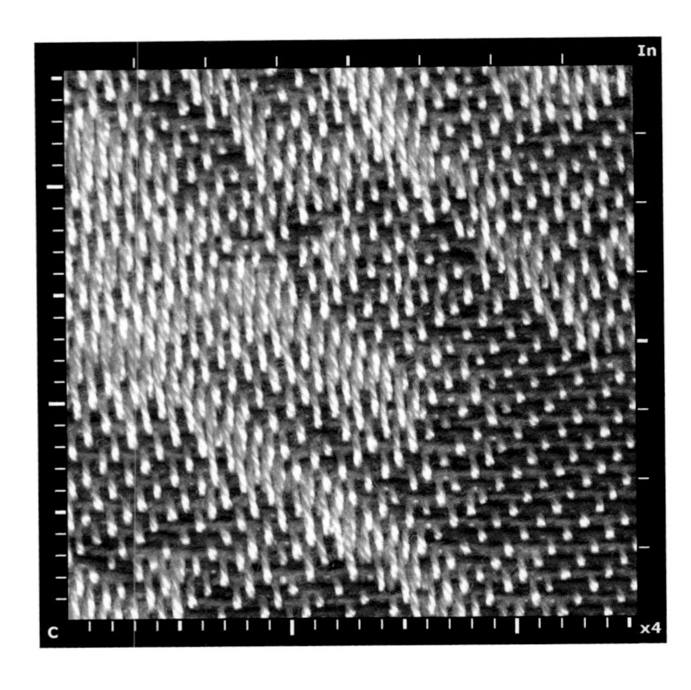

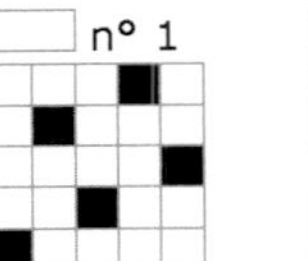

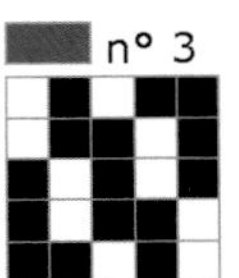

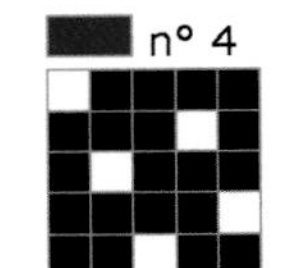

Figure 9.27. *Weaves and pointpaper design for* Frost; *detail.* Facing page: *photograph of* Skate Marks *and a snowflake chandelier inside* Warming Hut, *by Robin Muller. Courtesy of RM.*

Weave-patterned shibori

Woven shibori is the process of weaving supplementary pattern threads on a ground cloth, which are used to gather the fabric to form a resist for dyeing. It was inspired by nui or stitched shibori, a traditional technique practiced in Japan using a needle and thread. When the loom is used as the tool to place and organize the supplemental threads it is possible to achieve patterns that directly relate to woven structures. The Jacquard loom expands that vocabulary even more, making it possible to achieve large scale patterns and endless variations with or without repeats. Jacquard woven shibori is an opportunity to explore the intersection between weave structure, graphic design, and dyeing.

Catharine Ellis, textile artist and educator

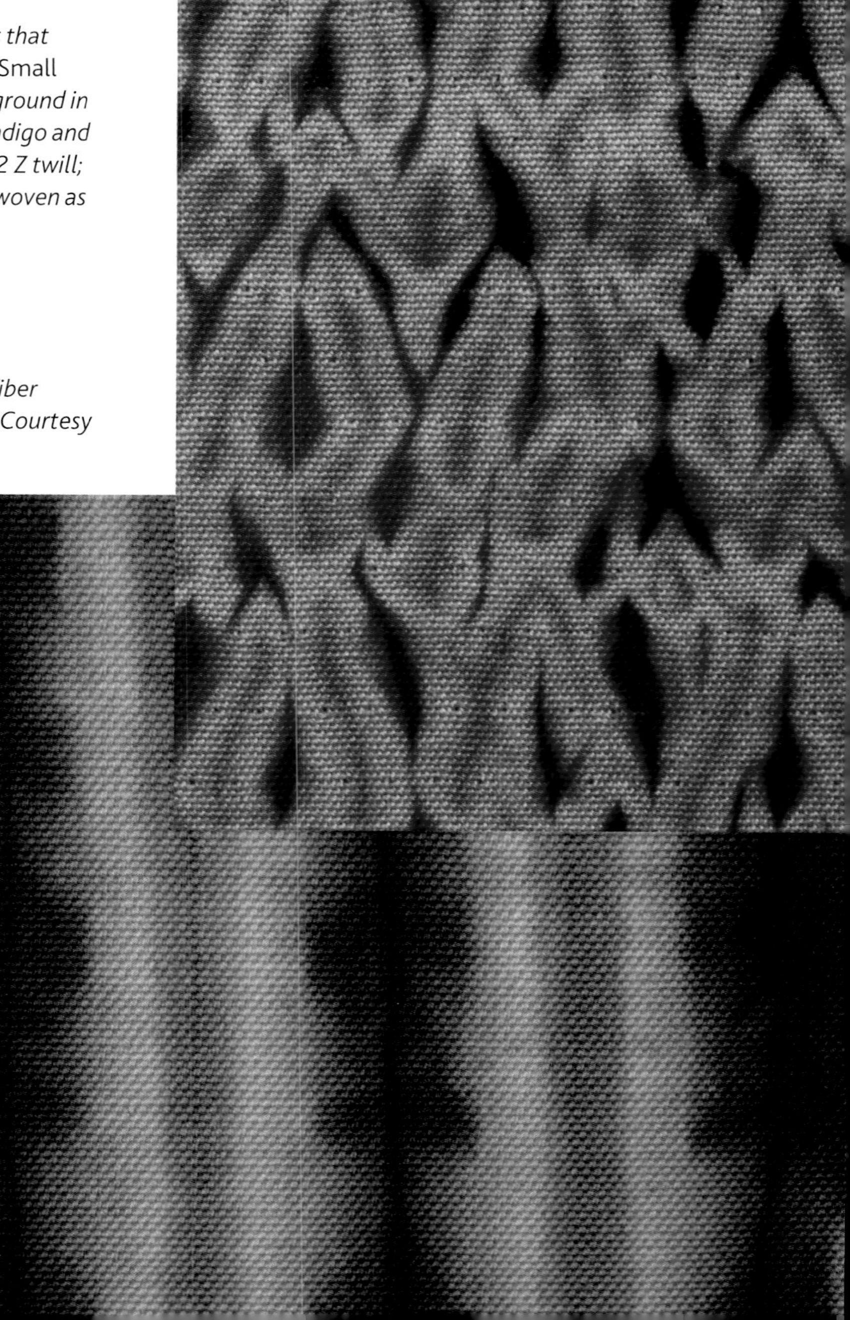

Figure 9.28. *Three shibori works by Catharine Ellis that incorporate weave patterning.* From left to right: Small All-Over Pattern, *cotton dyed with madder root, ground in plain weave;* Stripe Variations, *cotton dyed with indigo and quebracho black, ground woven as double-end 2/2 Z twill;* Big Circle, *cotton dyed with madder root, ground woven as double-end 2/2 reversing twill. Courtesy of CE.*

Figure 9.29. Below: Big Stripe, *cotton dyed with fiber reactive and vat dyes, ground woven using 1/3 rib. Courtesy of CE.*

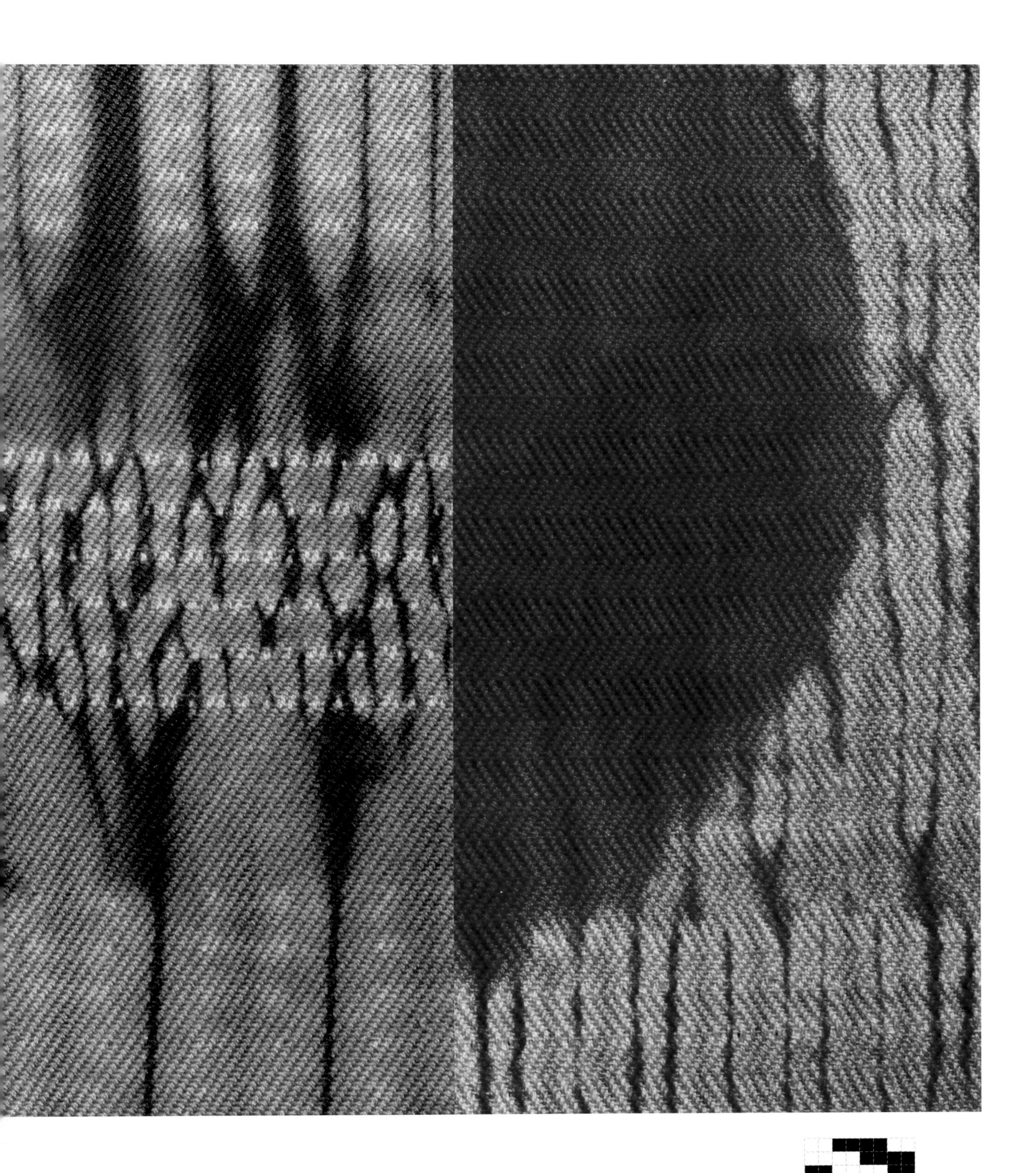

Figure 9.30. Right: *Ground weaves,* from left to right, *for the works in Figure 9.28:* Small All-Over Pattern, Stripe Variations, Big Circle.

Plain

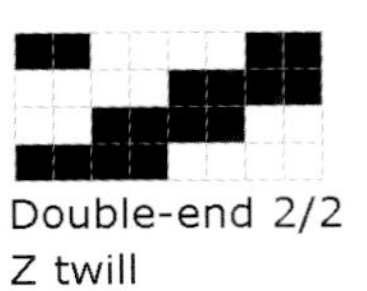

Double-end 2/2 Z twill

Reversed double-end 2/2 Z twill

Two compound structures, weave effects 1 and 2 pictured in Figure 9.32, are used to weave *Big Circle*. The ground weave in both effects is identical and woven with the first weft series. After ten ground picks, one pick of a second weft is inserted. In weave effect 1, the warp is raised above the supplementary weft; in effect 2, the warp rests below. Alternation between the two weave effects creates stitches across the width of the fabric. The interval between supplementary picks is regular—one pick every ten ground picks—but length and position of stitches vary across the fabric's width. No stitching occurs in the large circle at the center of the cloth.

When the supplementary wefts are pulled and knotted, vertical pleats form wherever stitching occurs. The unstitched circle at the center of the cloth remains unpleated. As the pleated and tied textile is immersed in the madder dye bath, only the edges of pleats and unstitched areas are free to absorb the dye; pressure exerted by the extra wefts hinders the dye from penetrating the pleated areas. Once dyeing is completed and the extra threads are removed, patterning becomes apparent. The regular interval between supplementary picks, combined with variations in stitch length and position, generate a rippled effect, whereas the absence of stitching results in a circular void, filled by the madder dye.

Figure 9.31. Top: Big Circle, *weave-patterned ground with extra wefts, before gathering and dyeing;* bottom: Big Circle *at real size with supplementary wefts gathered to create resist areas prior to dyeing. Courtesy of CE.*

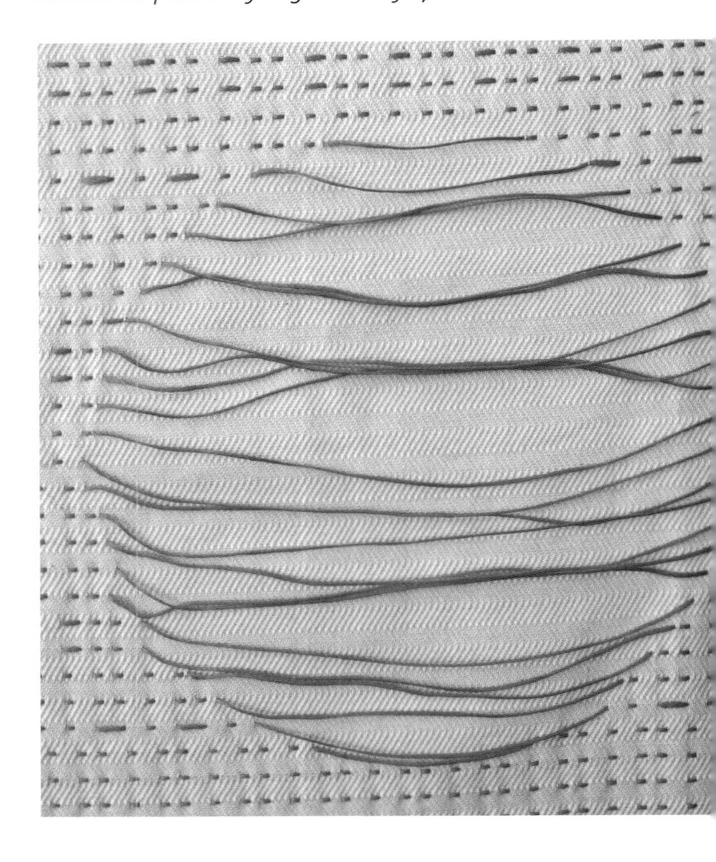

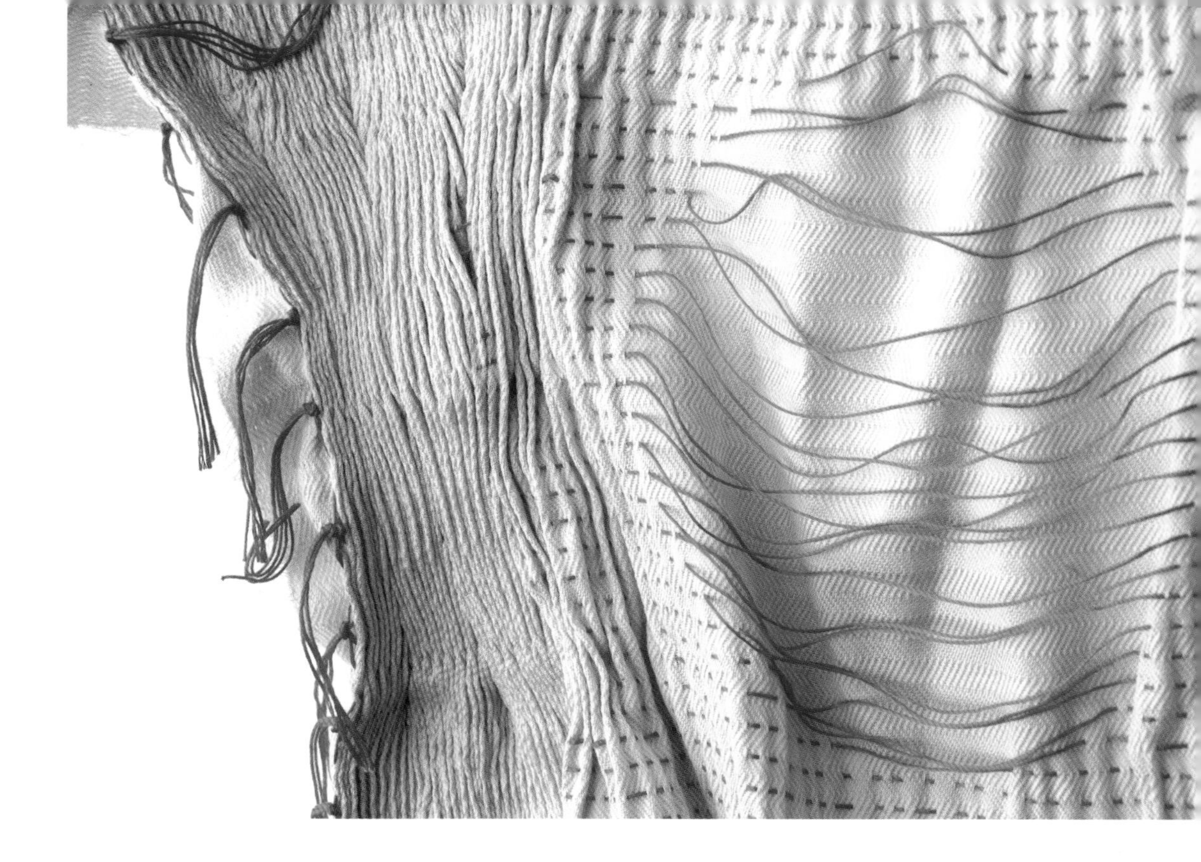

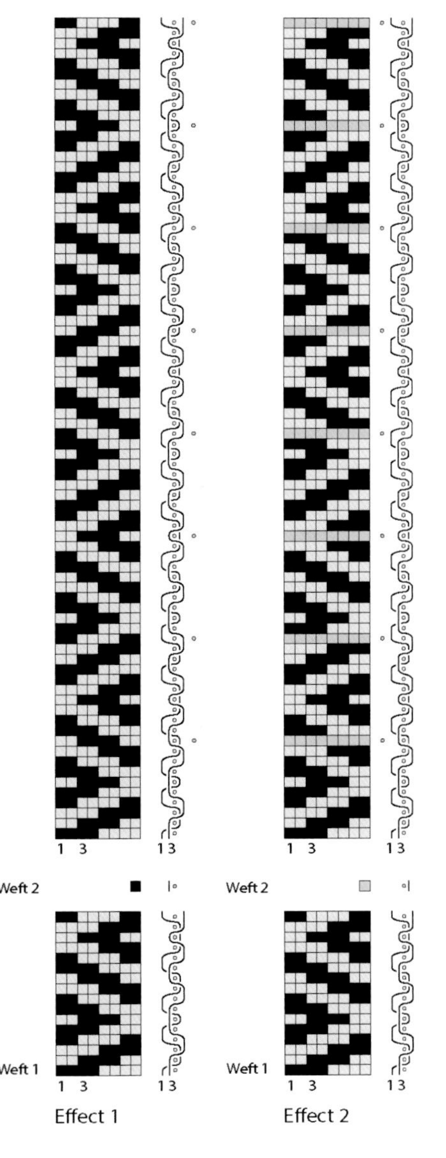

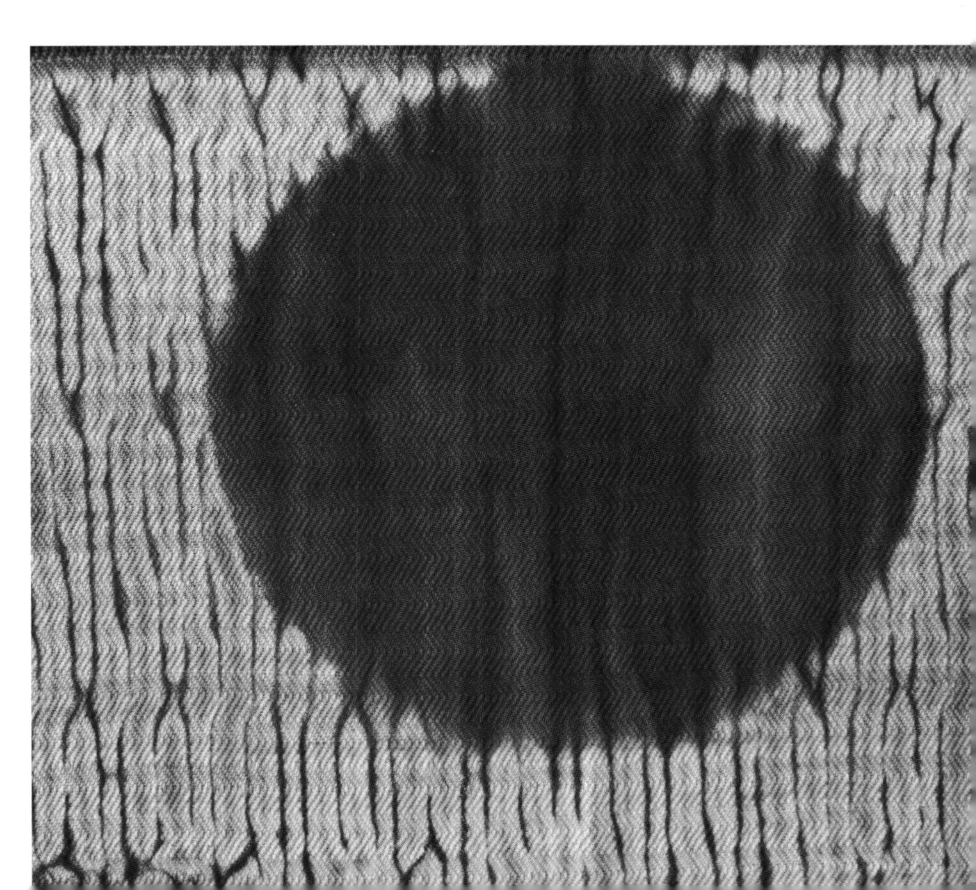

Figure 9.32. Above: *undyed work with pulled supplementary wefts;* below: *finished work after dyeing and removal of extra wefts;* center: *drafts showing the two compound structures used to weave* Big Circle. *Photographs courtesy of CE.*

Textiles in space and time

We think of the cloth of our everyday lives as bidimensional and unchanging. Elaine Ng Yan Ling's research focuses on textiles that contribute in functional or aesthetic ways to buildings in an urban context, where man's contact with nature and its cycles is reduced. Through use of imagery, structure, and materials, she creates interactive, changeable textiles that acquire and lose dimensionality, thus modifying light, color, and surfaces within work and living spaces. Modifications are transitory, progressive, or permanent; changes are rapid, seasonal, or require years; materials include today's programmable alloys, polymers, or natural materials that dilate/retract, release, or enmesh.

The sources, materials, and structures of Ng's work are exemplified by two projects from her Alive Furnishing Collection: *SuperLUX1* and *SuperLUX2*.

My design principle is based on *Biomimicry* and focuses on hybrid materialization of craft and technology. I use woven and etched patterns that respond to changes in environmental conditions such as light intensity or mechanical force.

Nature's endlessly evolving patterns—environmentally responsive tectonic structures—can be studied so that the design can both mimic and create new structures. This investigation explores the potential of nature's sensing system with the characteristics of shape memory alloy/polymers to create functional responsive structures and surfaces, and considers how the current construction system can be improved to suit a future, more efficient way of living.

I began the project by observing the shape memory behavior of living things. I saw how the behavior of wood has a natural engineering ability of movement. This inspired me to start to observe the tree itself, before it was cut into a thin sheet of wood veneer.

Within the city, trees are often grown within a confined space, forcing them to grow aerial roots. I then compared the roots to the cortical folds of the human brain, which also grow within a confined space. The motif for *SuperLUX1* derives from these observations, and is a 2d interpretation of 3d movement within nature.

Elaine Ng Yan Ling

Three wefts—1) silver polyester-metallic yarn, 2) copper wire, and 3) red spun polyester—are used to weave the three compound structures shown in Figure 9.35. In effects 1 and 3, the white polyester warp is split into two equal series to weave two layers of weft satin 8 with the silver weft 1 and red weft 3, respectively. In effect 1, the copper weft floats unbound between the two layers, in effect 3 the same weft floats unbound directly above a weft satin of the red weft, while the silver weft binds below in weft satin 8. In effect 2, the copper wire binds tightly with all warp ends in plain weave, while wefts 1 and 3 float unbound on the reverse.

Figure 9.33. *Aerial tree roots, Hong Kong. Courtesy of EN.*

Ng wove with copper wire for its reflective quality and to provide stability in a cloth woven of otherwise soft and slippery materials. "For the weft I chose copper wire; it is light sensitive in the sense that when seen from different angles, the wire reflects the different colors of its surroundings; placed on top of a red polyester yarn, it provides a multi-tone effect."

In narrow areas corresponding to effect number 3, the copper weft remains visible above the red weft; in wider

Figure 9.34. *Detail of light reflective properties of copper yarn visible in weave effects 2 and 3. Courtesy of EN.*

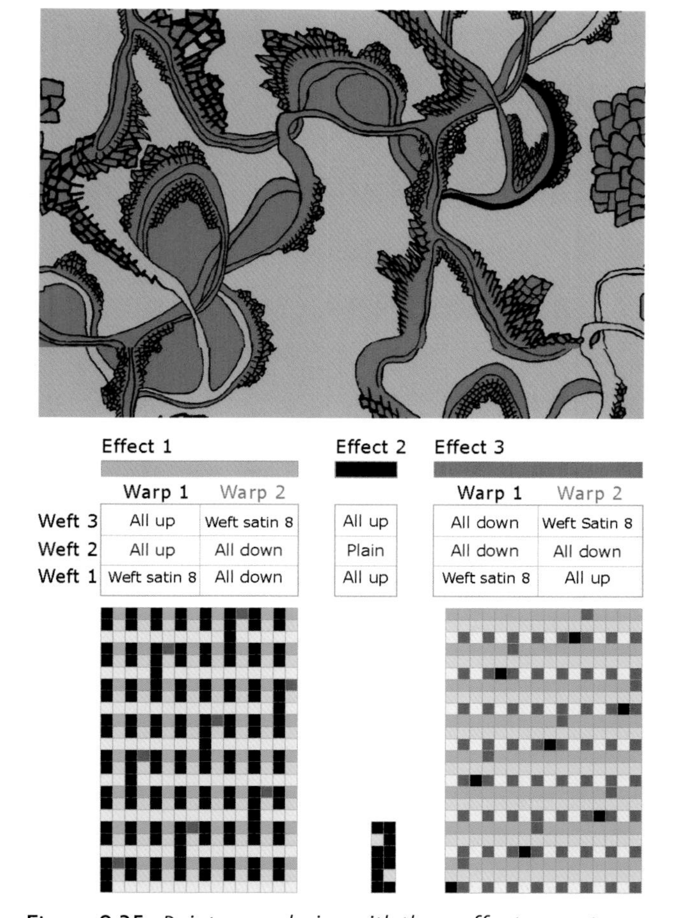

	Effect 1		Effect 2	Effect 3	
	Warp 1	Warp 2		Warp 1	Warp 2
Weft 3	All up	Weft satin 8	All up	All down	Weft Satin 8
Weft 2	All up	All down	Plain	All down	All down
Weft 1	Weft satin 8	All down	All up	Weft satin 8	All up

Figure 9.35. *Pointpaper design with three effects, courtesy of EN. Reading note.*

areas, Ng clipped away the metal wire to reveal the red weft satin beneath. In the next step of her process, she cut along the perimeter of the exposed red layer, leaving it attached to the body of the cloth by a small strip. Ng then embroidered a shape memory wire to the edges of the flaps with red thread.

Figure 9.36. *Detail of detached flap with shape memory wire application. Courtesy of EN.*

SuperLux1 is intended for installation in the public area of a hotel. When mounted on the wall of a foyer, movement sensors are placed on the exterior surface of the wall and connected to an Arduino system (a programmable microcontroller), which in turn is connected to an electrical circuit attached to the back of the fabric. When guests walk past the outer face of the wall, their movements are sensed and transmitted invisibly toward the interior, connecting the two spaces via an intelligent system that produces dimensional changes in the fabric. These modifications are transitory and rapid, and while triggered by movement outside the wall, are a reinterpretation of the growth patterns of tree roots.

Figure 9.37. SuperLUX1, *Jacquard woven multiple cloth with shape memory alloy and circuit, 36 cm x 110 cm / 14" x 43", for installation on the inside wall of a public area. Courtesy of EN.*

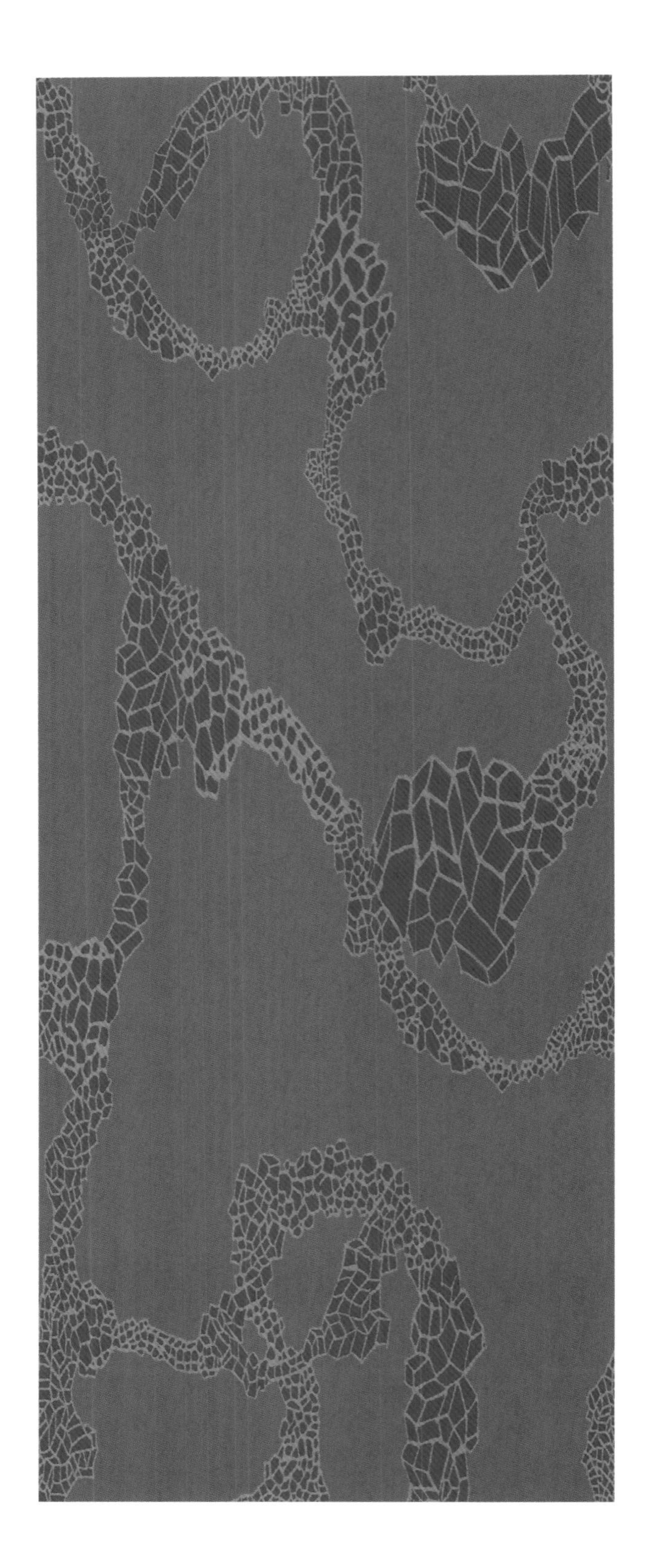

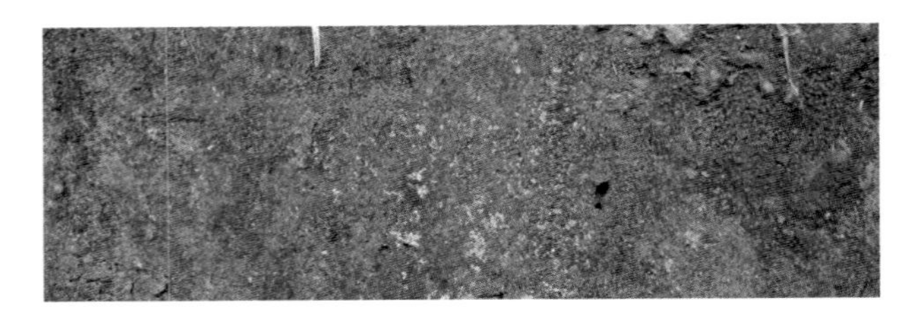

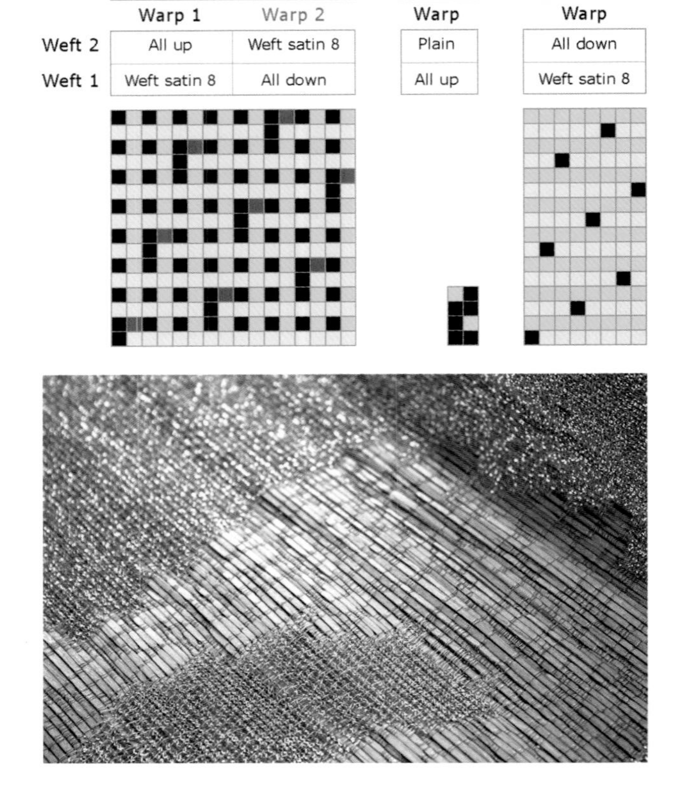

Figure 9.38. Above: *pointpaper with three weave effects for SuperLux2;* right, from top: *image of design source, weaves, detail of finished cloth. Courtesy of EN.*

SuperLux2 shares a number of elements with the previous project. Ng's objective was once again to produce a textile that could introduce movement and change into an urban space, such as the public area of a hotel. The motifs of the pointpaper design are inspired by nature in an urban context—the lichen-decorated bark of a tree in London. As in the previous work, the patterning in her design recalls a network of nerve cells.

Of the three weave effects, the ground weave, effect 1, is a weft satin doublecloth, woven with a white polyester warp. Unlike *SuperLux1,* only two wefts are used: the first is a highly spun silk/silver yarn, the second, thin strips of balsa wood. In weave effect 2, the silk weft floats unbound on the reverse, while all the warp ends bind the balsa weft in plain weave. In the third effect, the silk/silver weft weaves in weft satin 8 with all ends, while the balsa weft floats unbound on the face.

The first weft series weaves as weft satin 8 in both effects. The energy required to overspin the silk and silver weft was "captured" by wetting the spun yarn and then weighting it until dry. The second weft consists of thin strips of balsa, a less flexible material that provides stability and structure to the fabric, lending it a matt-like quality.

After weaving, the captured twist of the silk and silver weft was released when the designer steamed her textile to produce permanent shortening of the overspun weft. The balsa wood strips resist the silk's contraction and remain flat.

As seasons alternate throughout the year, the balsa wood contributes slow and reversible changes to the cloth's surface, swelling and contracting in response to variations in humidity and temperature.

Figure 9.39. *Detail and finished textile,* SuperLux2, *woven Jacquard double cloth with polyester warp, silver and silk weft, and balsa wood weft. Courtesy of EN.*

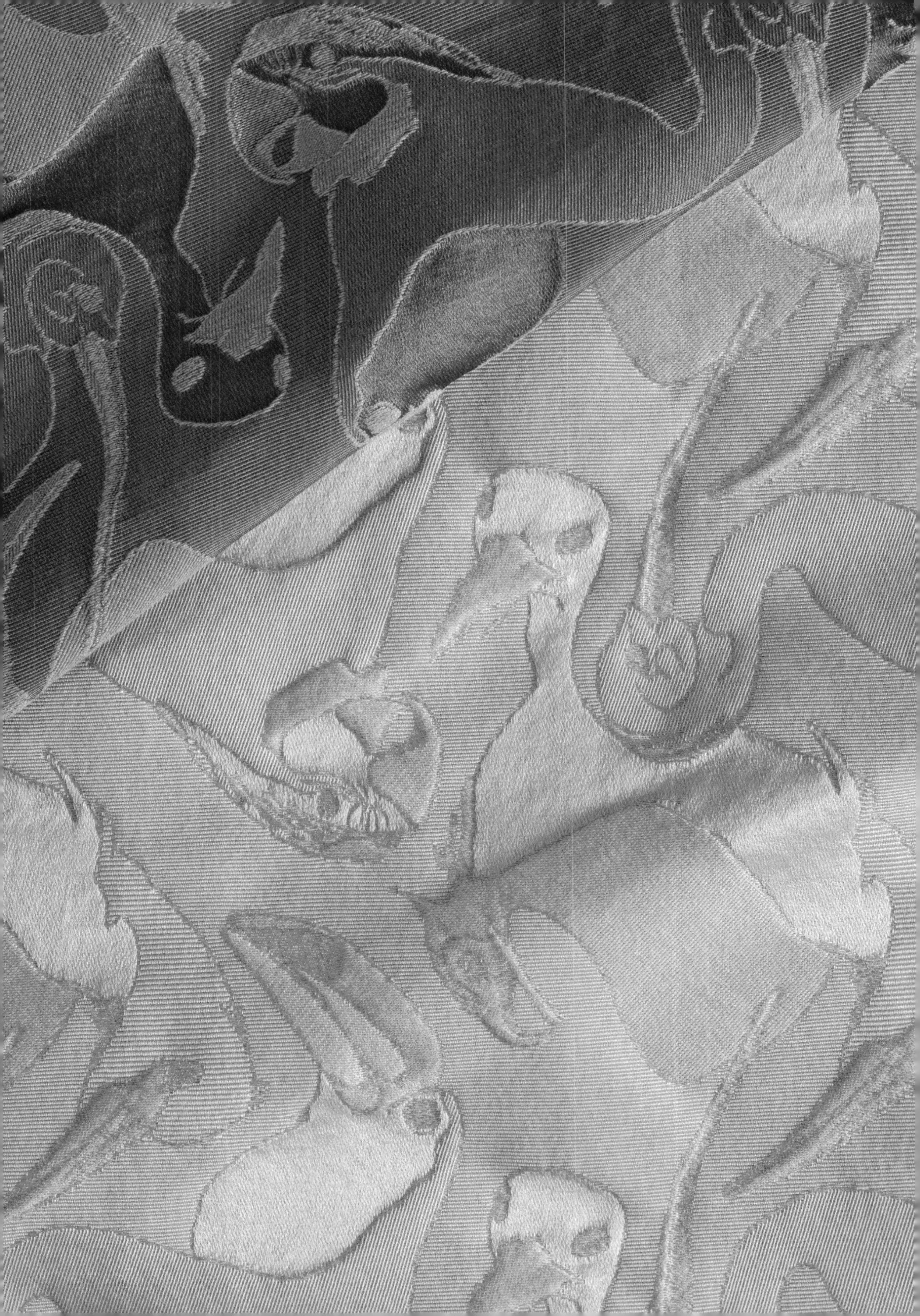

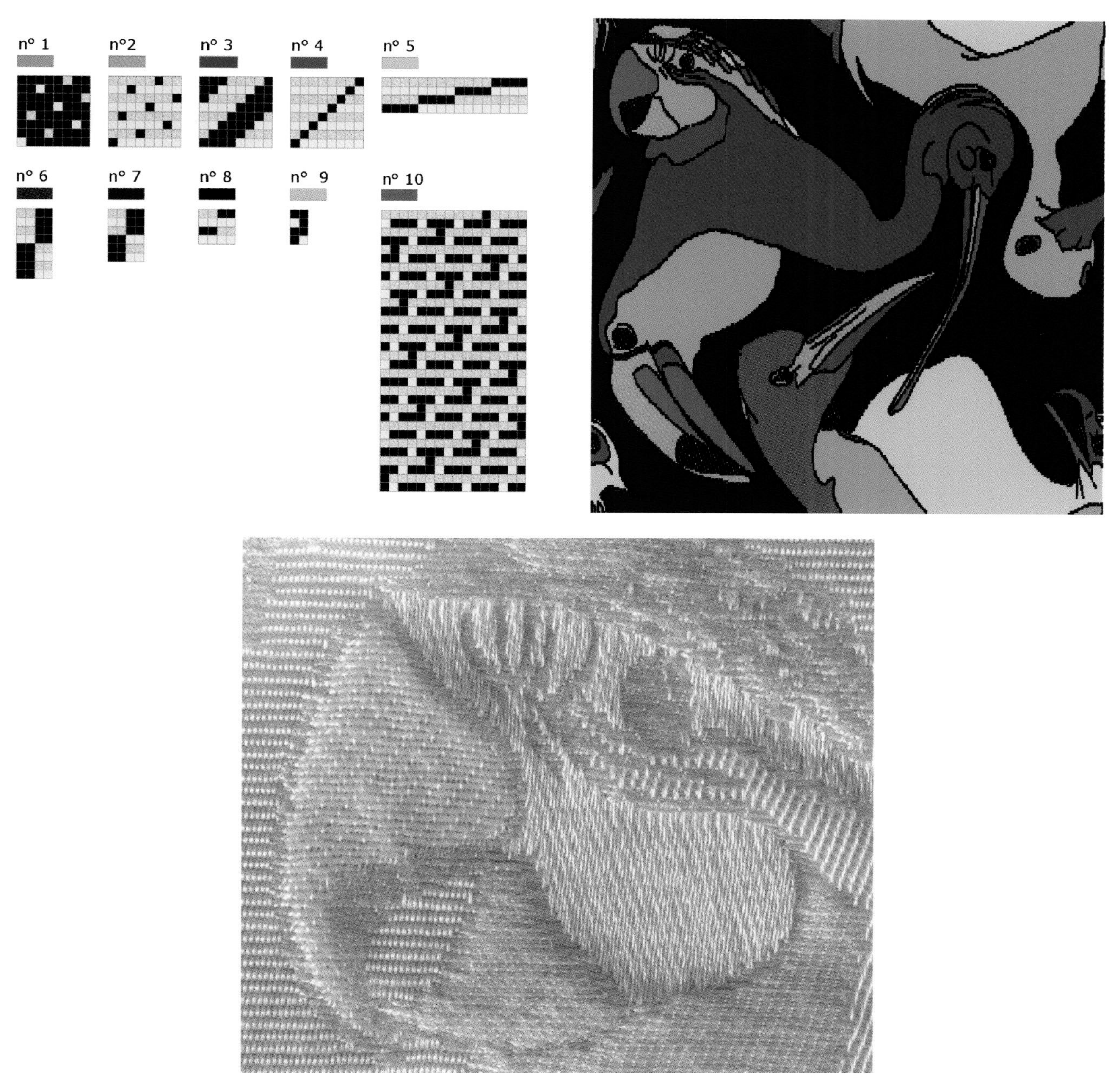

Figure 9.40. Facing page: Uccelli, *industrially woven silk and cotton damask with self-patterning wefts, by Daniela Tirabasso, LFS.* This page: *weaves and pointpaper, detail.*

How many weave effects?

To be effective, each weave structure in a figured textile must be distinct from all other effects in the same cloth. As the number of effects increases, the designer's task becomes more challenging, whether simple or compound weaves are used.

Uccelli, a damask with self-patterning wefts, contains ten weave effects and is woven with a single warp series. Two alternating ground weft series bind with the warp as a single weft in weave effects 1 to 7; in weave effect 8, the first weft series floats on the face, while the second weft binds in 1/2 rib to create the bed; weave effect 9 corresponds to binding points that cut overly long floats of weft one; in effect 10, the second weft is bound in weft satin 16, while the first weft weaves as a 1/2 rib below.

Small variations in the seven damask effects distinguish each simple weave; the longer and shorter weft floats in effects 8 and 10, whether contrasting in color or white and light beige, describe the other two distinct weave effects.

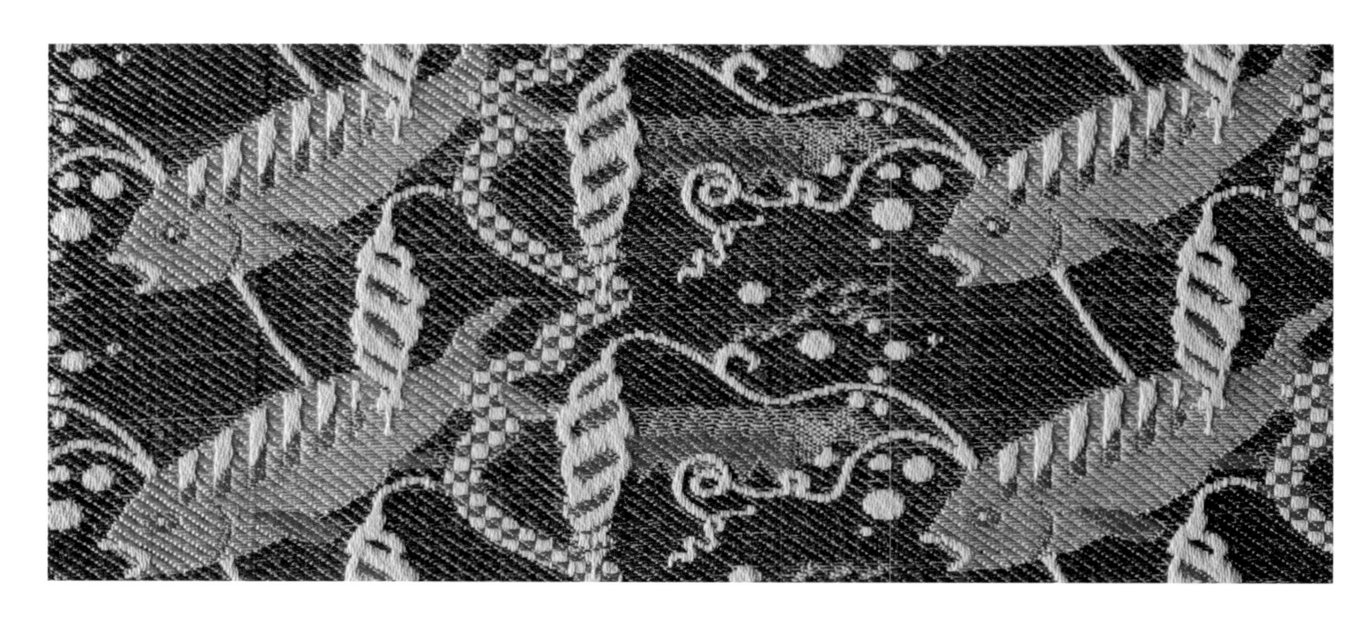

Once and future figuring techniques

Over time, design and patterning techniques fall into disuse, while others come into vogue, as weave-patterning technology and materials evolve. The next section of this chapter is dedicated to the use of patterning techniques from various eras, some forgotten, others in use, all interesting for today's designer.

Six Traditional Lampas

After lampas first appeared in Central Asia during the eleventh century,[1] it gradually replaced the earlier patterning techniques of taqueté and samitum. As opposed to its predecessors, which are reversible and weft faced on both sides, the unequal proportion of ground and binding warps produces two unequal faces and the warp becomes a visible patterning element. Weaving time is reduced, and the binding warp facilitates patterning with multiple wefts. See Chapter 6.

Figure 9.42. Chameleon *by Eva Schlechte, warp satin ground, silk and metallic wefts bound in 1/3 S twill, LFS/Stuttgart Academy.*

Figure 9.41. Facing page, row by row: Briciola at Night *by Barbara Shawcroft, warp satin ground, pattern and brocading wefts bound in 1/3 Z twill, LFS;* Buttoned Boots *by Heidemarie Honenbüchler, damask ground with silk, silver and gold wefts bound in 1/3 S twill, LFS; eighteenth-century lampas with tabby ground, silk pattern and brocading wefts bound in 1/3 Z twill, courtesy of Cora Ginsburg LLC, NYC; eighteenth-century lampas with self-patterning, pattern, and brocading wefts bound in 1/3 S twill, courtesy of Societé Le Manach;* Simon's Fish *by Simon Peers, warp satin ground with silk pattern wefts bound in 1/3 S twill, LFS. NB: The term* ground *refers not to the ground effect, but to the ground weave, which may appear on the face or lie beneath weft patterning.*

The six figured textiles on these two pages are traditional lampas woven on looms with a ground warp mounted on both a figuring and ground harness, and a second binding warp controlled by yet another harness. The binding warp binds all pattern wefts with the same weave structure, whether these lie above or below the ground warp and weft. The ground may be simple, or figured in its own right, as in the lampas *Buttoned Boots,* or the eighteenth-century lampas with a yellow silk ground directly below.

Figure 9.43. Morning, *doublecloth lampas in silk, paper, and cotton by Tanja Valta, LFS;* facing page*: digital pointpaper; detail* (top) *and drafts* (bottom) *of each weave effect.*

Morning is a doublecloth, damask, and a lampas woven with one ground weft and one pattern weft. While ground warp and weft weave as simple structures to create warp to weft-faced surfaces typical of damask, the binding warp weaves a separate layer of plain weave with the pattern weft on the reverse in all effects but one. Only in effect 1 does the pattern weft appear above the ground warp and weft, tightly bound in plain weave by the binding warp.

Contemporary in design, *Morning* combines figuring technology, design, and the woven medium in ways shared with the two *Bizarre* silks on the next pair of facing pages: negative and positive spaces, ground and patterning areas create abstract forms. Stylized, fantastic flora is transformed into woven surface with shifting colors, reflective and opaque materials. Figurative, but abstract, all three textiles use the unique language of woven cloth in a way that cannot be replicated by textile mediums such as print, knit, quilting, and embroidery, nor by the two-dimensional mediums of painting, print, or photography.

Figure 9.44. *On the pages that follow, two eighteenth-century* Bizarre *silks with metallic brocading wefts.* Left page: *courtesy of Cora Ginsburg LLC, NYC;* right page: *courtesy of Societé Le Manach.*

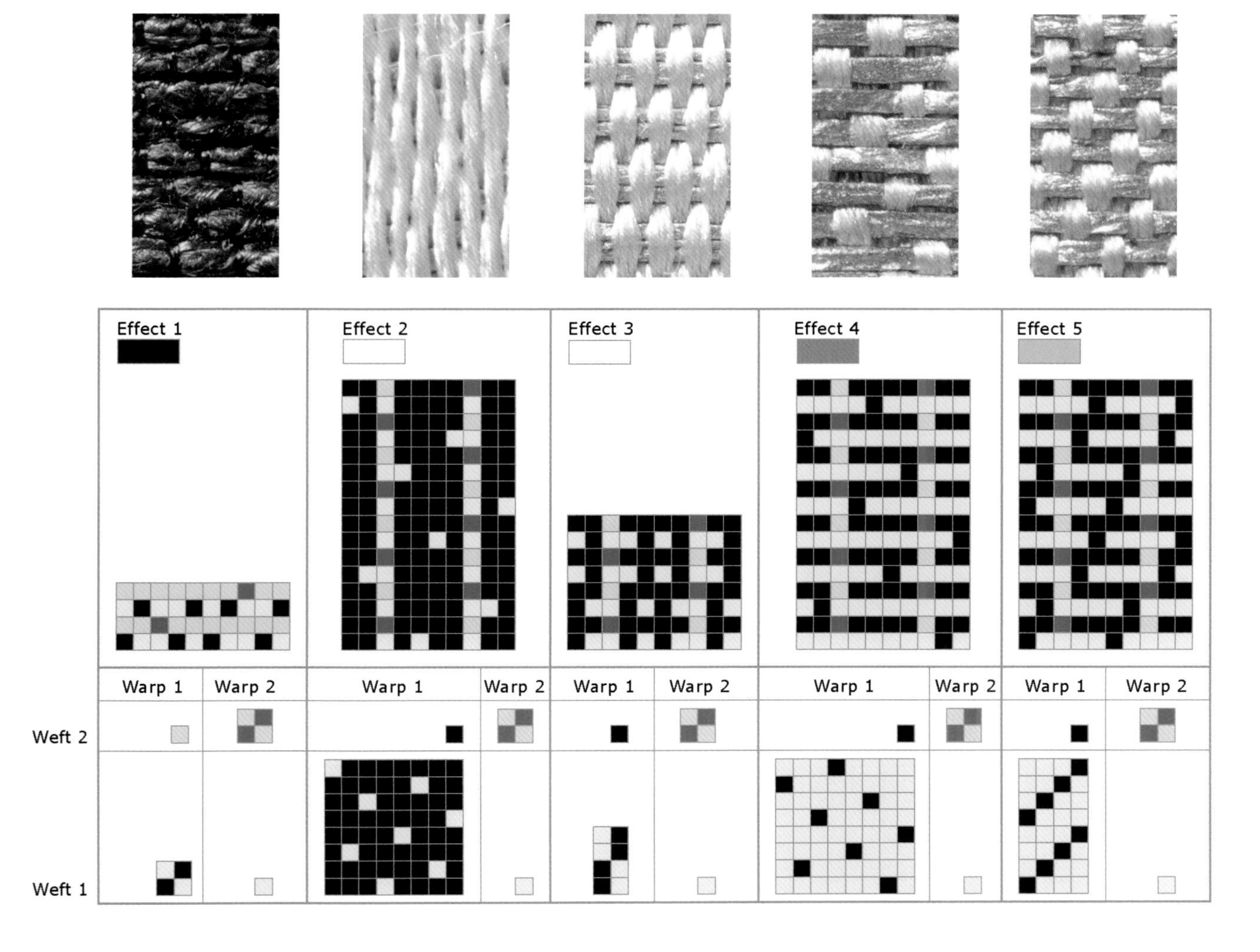

Economical Use of Warp and Weft Color

Introduced in the 1730s,[2] *point rentrée* rapidly came into vogue as a means of increasing the number of perceived colors on a textile's surface with no additional material or weaving time. A simple device uses two weft colors to produce a third, intermediate shade. Every other pick of a first weft series alternates with every other pick of a second weft series. In the past, this device was drawn on the pointpaper as alternating bars of the two colors representing the two weft colors, as shown in the uppermost portion of Figure 9.45. Today, this blended effect may be easily drawn on a digital pointpaper with a third color, which is then defined as "warp down" when weave structures are assigned to alternate picks of the two wefts colors.

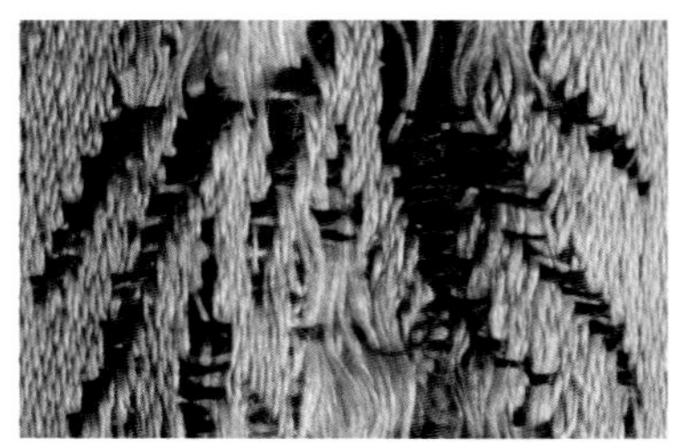

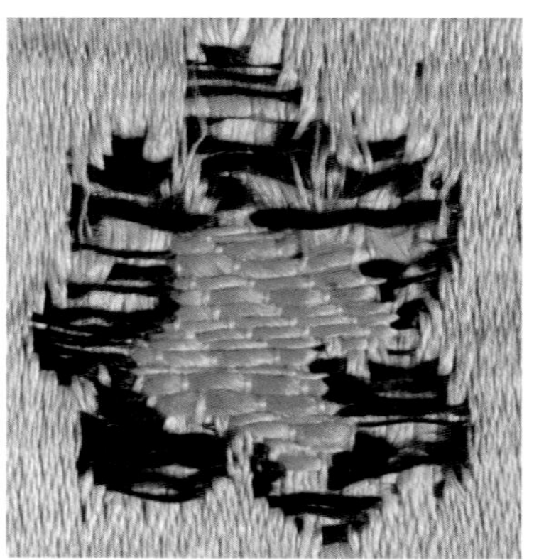

Figure 9.45. Point rentrée *as represented in the past, or as one would today, with a distinct color on the pointpaper; point rentrée patterning. Courtesy of Societé Le Manach.*

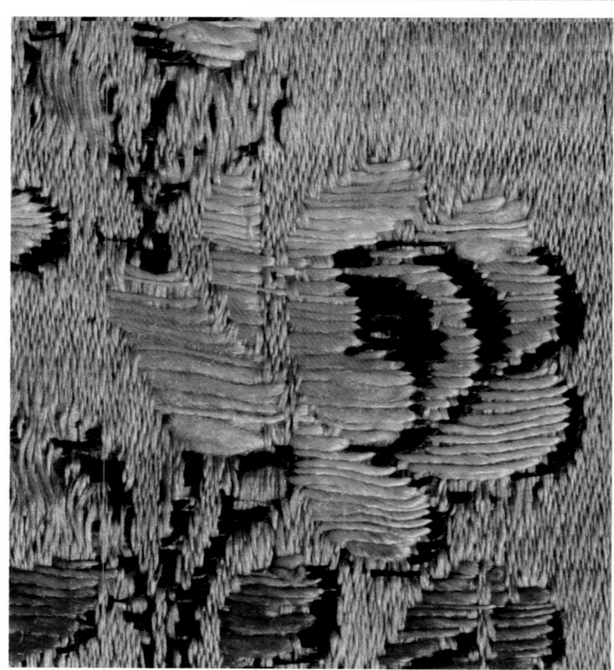

Figure 9.46. *Eighteenth-century figured satin with striped ground warp and multicolored self-patterning wefts. Courtesy of Societé Le Manach.*

Another device used to augment patterning colors with no increase in material and weaving time is shown in figured satin below and on the facing page. One warp and two weft series produce a warp satin ground. A first weft series in solid black floats on the face where required to produce fine lines and the fur or feather motifs that soften the warp striping. Colors of the second weft series vary in correspondence with several floral motifs in yellow, pink, and green. A high-density warp satin ground, together with the unifying presence of the solid black weft series, effectively hides the color changes of the second weft series.

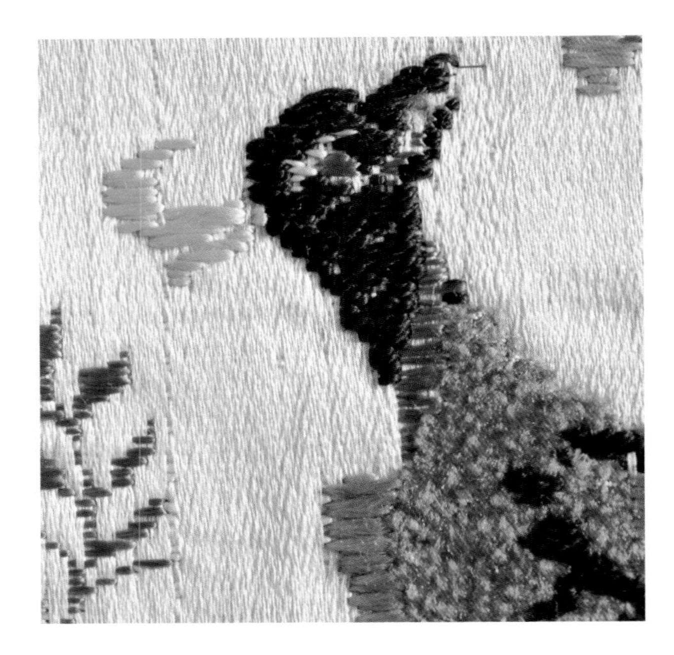
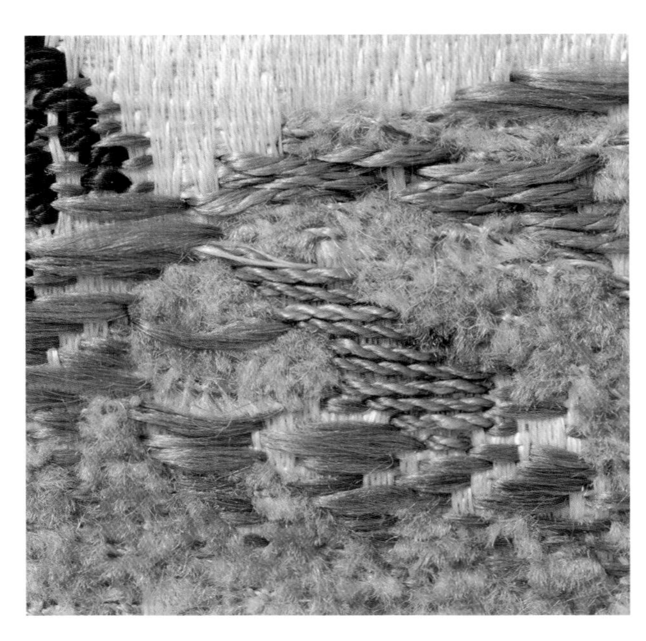

Figure 9.47. *Seventeenth-century brocaded satin, CJH.*

Brocading with an Abundance of Wefts

Inevitably brocading adds design and weaving time as well as material to a figured textile, but unlike pattern and ground weft patterning, which have an impact on the color and density of the ground, brocading is a "local" effect, limited to the area of the motif to be woven. Brocading has little to no structural impact on the overall textile, allowing the designer great freedom of design, choice of material, and color.

The small figures of birds and botanica that decorate the satin ground are brocaded with silk floss, cord, and chenille in white, green, blue, pale and bright yellow, pink, three shades of coral, and two of brown. In any one figure, weft line per weft line, two to three colors are present, requiring as many shuttles and separate sheds. Woven face down, this textile was produced slowly, and required a skilled weaver with a good memory for color, motif, and weft sequence.

Impossible to weave industrially, today the production of brocade is hugely facilitated by digital figuring devices mounted on handlooms, able to transmit the many sheds required by multiple weft series.

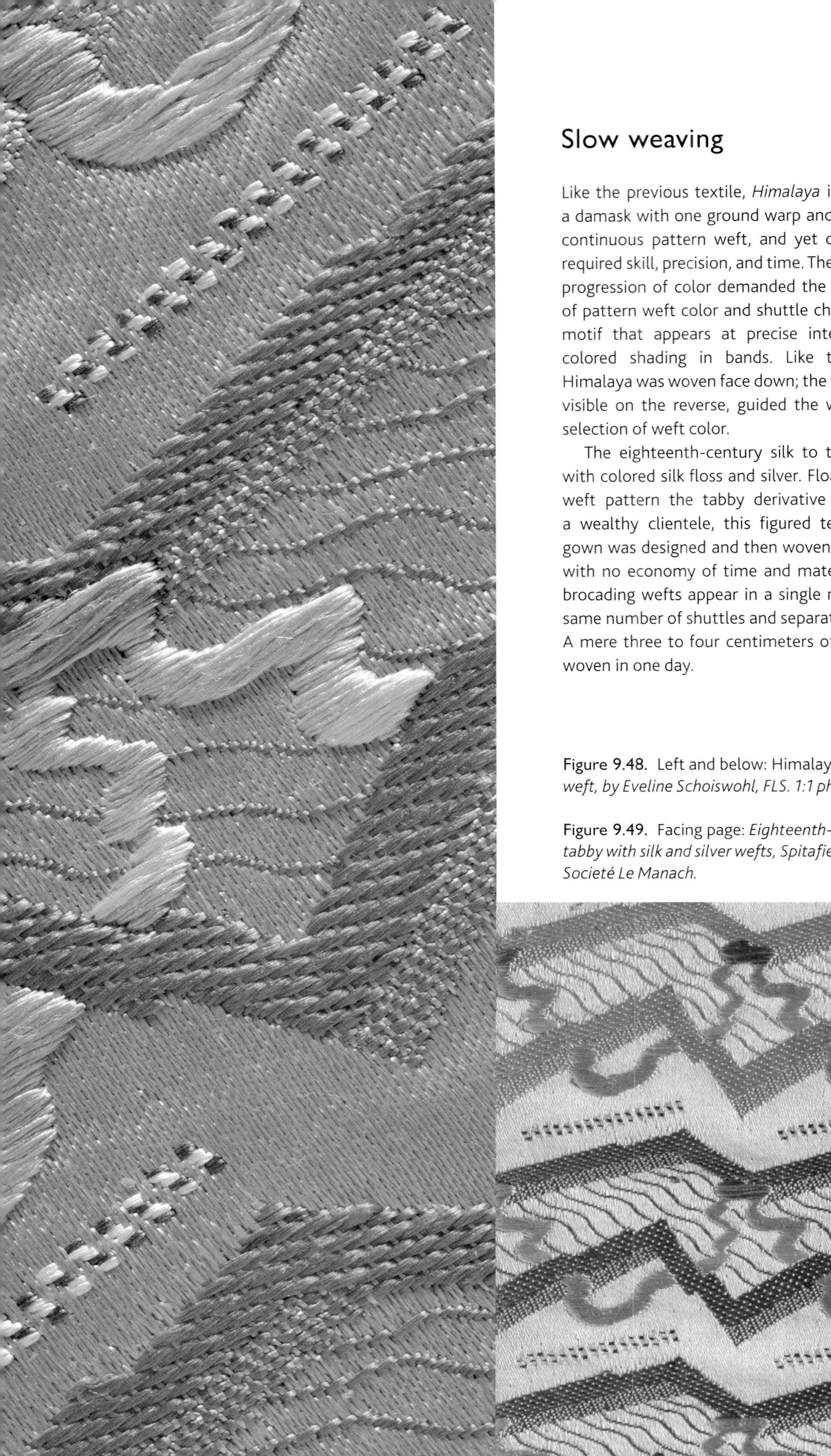

Slow weaving

Like the previous textile, *Himalaya* is structurally simple, a damask with one ground warp and weft series, and one continuous pattern weft, and yet design and execution required skill, precision, and time. The gradual rainbow-like progression of color demanded the constant adjustment of pattern weft color and shuttle changes. The checkered motif that appears at precise intervals interrupts the colored shading in bands. Like the previous textile, Himalaya was woven face down; the wide expanse of weft, visible on the reverse, guided the weaver in the careful selection of weft color.

The eighteenth-century silk to the right is brocaded with colored silk floss and silver. Floats of a white ground weft pattern the tabby derivative ground. Created for a wealthy clientele, this figured textile for a woman's gown was designed and then woven by skilled specialists, with no economy of time and material. As many as five brocading wefts appear in a single motif and require the same number of shuttles and separate sheds per weft line. A mere three to four centimeters of such a cloth can be woven in one day.

Figure 9.48. Left and below: Himalaya, *damask with pattern weft, by Eveline Schoiswohl, FLS. 1:1 photograph and detail.*

Figure 9.49. Facing page: *Eighteenth-century brocaded tabby with silk and silver wefts, Spitafields. Courtesy of Societé Le Manach.*

Piazza San Marco—Day

Velvet is perhaps the most challenging of all techniques to weave. Regularity of beat for ground and patterning sheds, the strength required to cinch the pile ends between picks of the ground weft, deft insertion of the fine velvet rods, and the sure and steady gesture required to cut the pile and free each successive rod are achieved over a long period of training.

Piazza San Marco—Day is one work of artist-educator and velvet expert Jan Paul. In this piece, she adds two rare effects to the ground, uncut and cut structures of ciselé velvet: brocade and alluciolato.

Brocade is generally woven face down to leave the unsightly turnaround points of the weft on the reverse, in the same manner that the less attractive stitches of embroidery remain on the cloth's reverse. Velvet may be brocaded from the face in areas where the ground; abuts the cut pile effect. When cut, the freed ends of the pile warp expand, providing a screen of silk filaments that hide the turning points of the gold brocading weft.

The second effect, *alluciolato* or *firefly velvet,* is a rarely executed technique for which the velvet weavers of Florence were renowned during the Renaissance. Visible between the tufts of cut pile, *alluciolato* is created by raising a short length of a gold brocading weft between pile ends with a hook to create the small circular loops that glimmer within the cut pile areas. Particular skill is required, and the creation of the loops slows weaving to a snail's pace.

Chiné Lace Velvet

Like *Piazza San Marco—Day,* the velvet pictured here is a rare and demanding work, the sum of multiple textile processes. The chiné warp is printed before weaving; the fine uncut pile decorates the ground with a large lace and flower motif. The pile warp is bound into the ground solely where it appears on the face of the textile. To conserve the light colors and hand of the velvet's ground, when not employed for figuring, the pile ends float unbound on the reverse; when weaving is completed, the floating pile is painstakingly cut away.

Figure 9.50. *Detail of* Piazza San Marco—Day, *ciselé, brocaded alluciolato velvet by Jan Paul, LFS; real size.*

Figure 9.51. Facing page: *Face and reverse of uncut, clipped end "lace" velvet on a chiné taffeta ground; mid-nineteenth century. Courtesy of Societé Le Manach.*

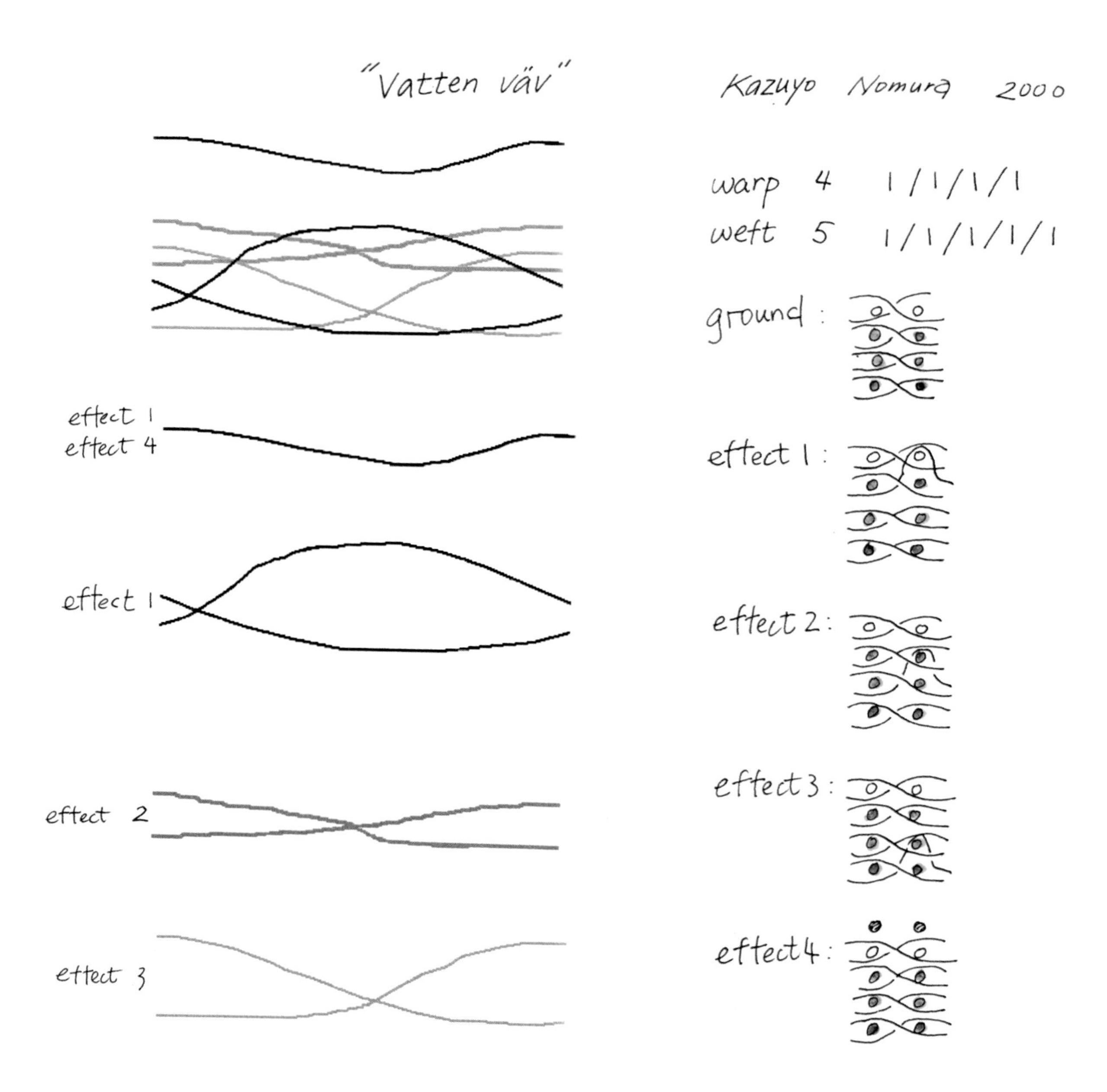

	ground 1 2 3 4	effect 1 1 2 3 4	effect 2 1 2 3 4	effect 3 1 2 3 4	effect 4 1 2 3 4
○	T D D D	T D D D	T D D D	T D D D	T D D D
●	U T D D	U U D D	U T D D	U T D D	U T D D
●	U U T D	U U T D	U U U D	U U U D	U U T D
●	U U U T	U U U T	U U U T	U U U T	U U U T
●	—	—	—	—	D D D D

Vatten Väv: Weaving Water

The design and compound weave structures of *Vatten Väv* were carefully pondered to generate the quasi invisible stitching that occurs between the equivalent warp and weft series of four plain weave layers. In the ground effect, all four layers remain separate. In effects 1, 2, and 3, the alternate ends of warp series 2, 3, and 4 rise upward one layer to bind the even picks of weft series 1, 2, and 3, respectively. In the last weave effect, a fifth and discontinuous weft series is shown above all layers. By comparing the detail of the finished sample, pointpaper, and weave information, the reader may trace the itinerary of the figuring wefts as these travel through the layers, length, and width of the repeat.

Though sober in materials and color, the weaving of this quadruple cloth with widely spaced figuring picks progressed as slowly as that of a richly patterned silk and gold brocade. The pattern wefts' paths through the weave defy our understanding and one of the constrictions of woven textiles: the perpendicular, consecutive interlacement of warp and weft. No break can be detected in any of these weft insertions, each woven with a single length of yarn.

Figure 9.52. Facing page: *pointpaper designs, sectional drafts, and reading note for* Vatten Väv, *a figured quadruple cloth with pattern weft by Kazuyo Nomura, LFS;* this page: *detail;* following pages: *finished work.*

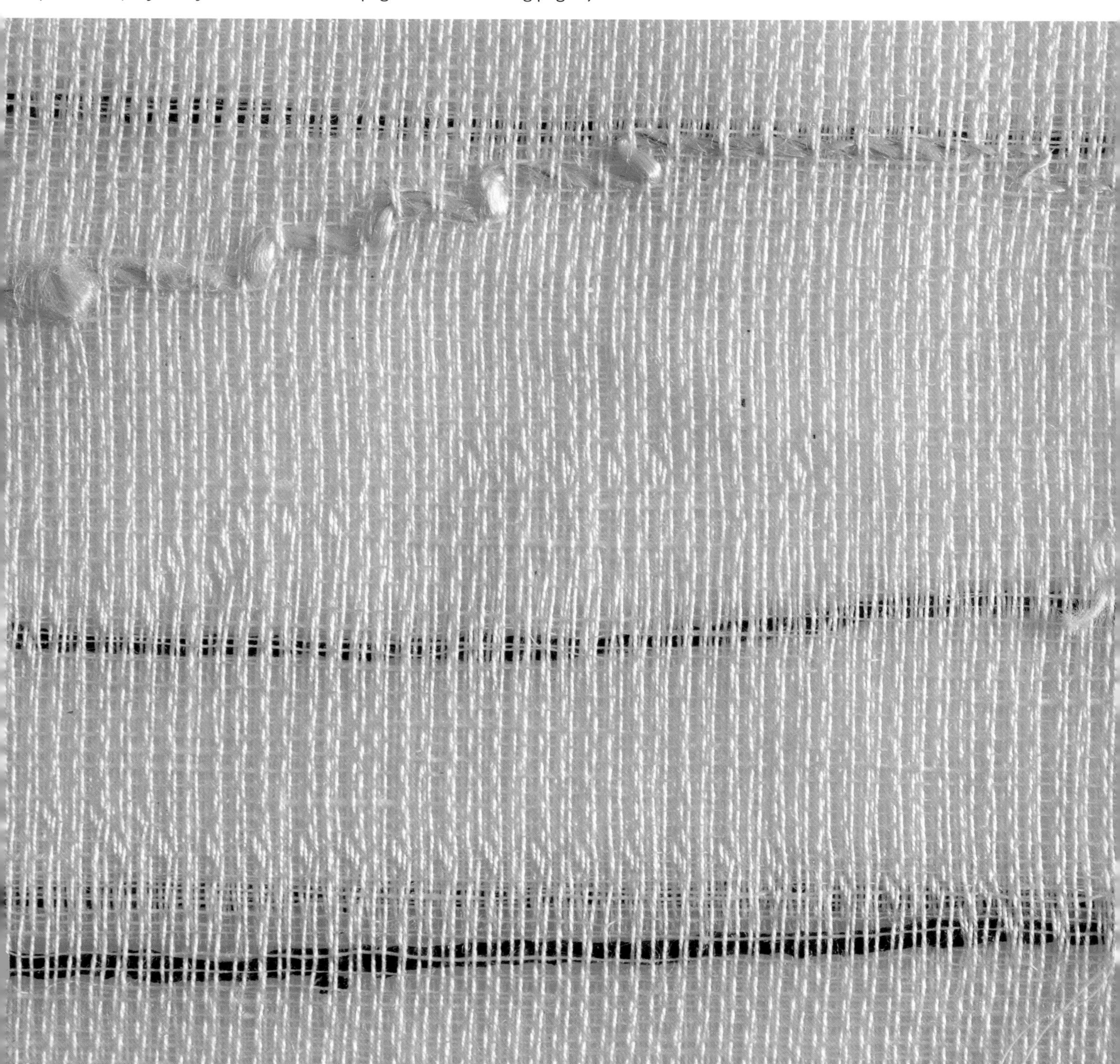

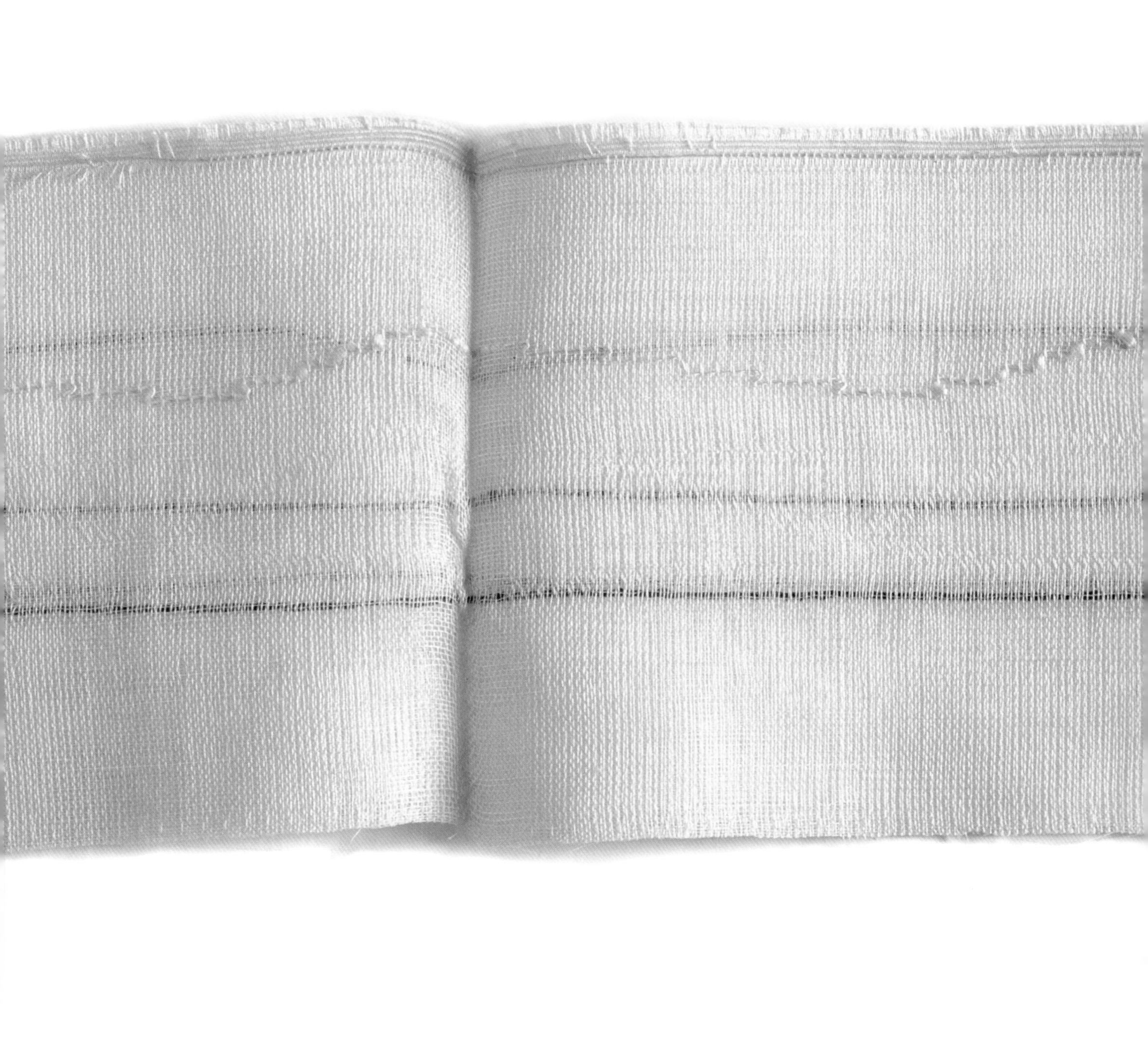

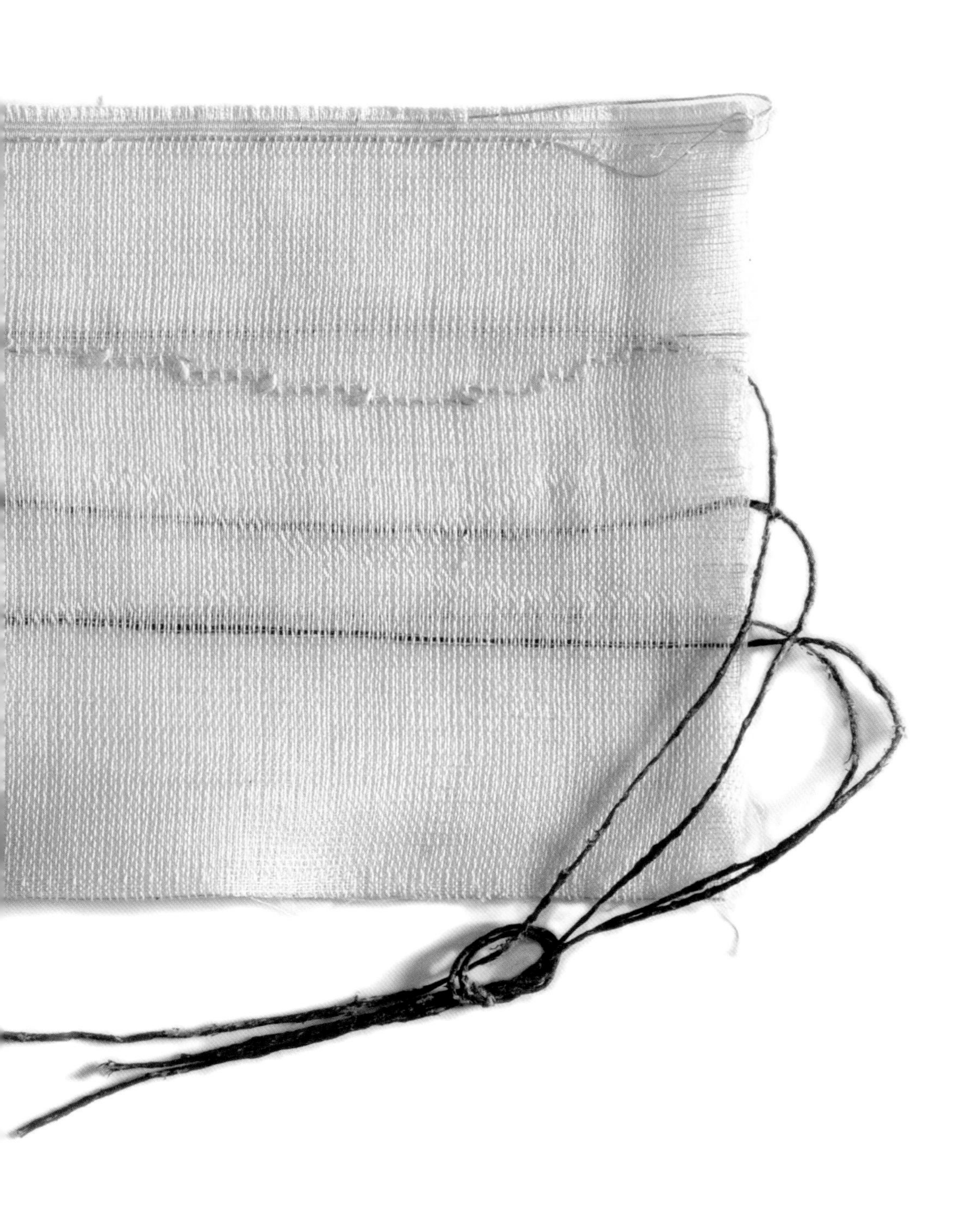

Notes

Introduction

1. A *Lucchese braccio* is 58.362 centimeters or 22.95 inches in length. For widths and costs per *braccio* of textiles produced in Italy during the Renaissance, see tables in Lisa Monnas, *Merchants, Princes and Painters* (New Haven, CT: Yale University Press, 2008), pp. 321–322.

Chapter 1: Visual analysis

1. Vilhelm Slomann, *Bizarre Designs in Silk* (Copenhagen: Munksgaard, 1953). Slomann establishes the use of the term *bizarre* to describe figured textiles with large, asymmetrical designs and exotic motifs found in European textiles from the late seventeenth to the early eighteenth centuries.

Chapter 4: Weave-patterning technology in the digital era

1. Feng Zhao, *Treasures in Silk* (Hong Kong: ISTAT Costume Squad, 1999), p. 39.

2. Agnes Geijer, *A History of Textile Art* (London: Philip Wilson Publishers, 1979), p. 137.

3. Ibid., p. 141.

Chapter 6: Figuring techniques

1. Feng Zhao, *Treasures in Silk* (Hong Kong: ISTAT Costume Squad, 1999), p. 334.

2. Agnes Geijer, *A History of Textile Art* (London: Philip Wilson Publishers, 1979), p. 137.

3. Ibid., p. 141.

Chapter 9: Case studies

1. John Becker, *Pattern and Loom: A Practical Study of the Development of Weaving Techniques in China, Western Asia and Europe* (Copenhagen: Rhodos, 1987), pp. 145–149.

2. Anna Jolly, *Seidengewebe des 18. Jahrhunderts II: naturalismus* (Bern: Abegg-Stiftung, 2002), p. 377: "The invention of the point rentrée or berclé effect is ascribed to Jean Revel (the famous Lyonnaise designer) at the beginning of the 1730s (a mise en carte with point rentrée is dated 1733)."

Bibliography

Becker, J., *Pattern and Loom: A Practical Study of the Development of Weaving Techniques in China, Western Asia and Europe,* Copenhagen: Rhodos, 1987.

Broudy, E., *The Book of Looms,* Lebanon, NH: New England Press, 1979.

Burnham, D. K., *A Textile Terminology, Warp and Weft,* London and Henley: Routledge & Kegan Paul, 1981.

Buss, C., *Seta, Oro e Argento,* Milan: Fabbri, 1992.

C.I.E.T.A, *Vocabulary of Technical Terms,* Lyon: C.I.E.T.A, 1964.

DeMoor, A., Verhecken-Lammens, C., and Verhecken, A., *3500 Years of Textile Art,* Tielt: Lannoo, 2008.

Diderot e D'Alembert, *L'Encyclopedie, Art de la Soie,* Paris: Bibliotheque de l'Image, 2002.

Emery, I., *The Primary Structures of Fabrics,* New York: The Textile Museum, 1966.

Geijer, A., *A History of Textile Art,* London: Philip Wilson Publishers, 1979.

Grosicki, Z. J., *Watson's Advanced Textile Design,* 4th ed., Cambridge: Woodhead, 2004.

Jolly, A., *Seidengewebe des 18. Jahrhunderts II: naturalismus,* Bern: Abegg-Stiftung, 2002.

Kolander, C., *A Silk Worker's Notebook,* Myrtle Creek, OR: published by author, 1979.

Monnas, L., *Merchants, Princes and Painters,* New Haven, CT: Yale University Press, 2008.

Pompas, R., *Textile Design,* Milan: Hoepli, 1994.

Puliti, M., *Disegno Tecnico Tessile,* Florence: Media, 1990.

Quinn, B., *Textile Futures,* Oxford: Berg, 2010.

Riboud, K., *Samit & Lampas,* Paris: A.E.D.T.A., 1998.

Schlein, A. and Ziek, B., *The Woven Pixel,* Greenville, SC: Bridgewater Press, 2010.

Sheehan, D. and Sutton, A., *Ideas in Weaving,* Loveland, CO: Interweave Press, 1989.

Slomann, V., *Bizarre Designs in Silk,* Copenhagen: Munksgaard, 1953.

Sutton, A., *The Structure of Weaving,* London: Hutchinson, 1982.

Van der Hoogt, M., *The Complete Book of Drafting for Handweavers,* Petaluma, CA: Unicorn Books and Crafts Inc, 2000.

Watts, J.C.Y. and Wardwell, A. E., *When Silk Was Gold,* New York: Metropolitan Museum of Art, 1997.

Weibel, A. C., *2000 Years of Silk Weaving,* New York: Weyhe, 1944.

Zhao, F., *Treasures in Silk,* Hong Kong: ISTAT Costume Squad, 1999.

Contributors

Individuals

Andéer, Tove *Reindeer, Tove's velvet*

Araldo, Anna *Rose*

Bajardi, Morgan *Facett*

Ciszuk, Martin *Haitienne, Liljeblom, Martin's Droguet*

Conly, Gregg *Bauhaus*

Crociani, Cecilia *Nettles*

d'Ambrosio, Carmel *Grass*

Egger, Elisabeth *Green Triangles*

Ellis, Catharine *Big Circle, Big Stripe, Small All-Over Pattern, Stripe Variations*

Fernström, Päivi *Päivi's Velvet, Pomegranate*

Forchhammer, Berthe *Bolsterlang, Fragment*

Golestaneh, Vanessa *Vanessa I*

Haglund, Emelia *Grenade*

Hansen, Anne-Birgitte *Fingerprint*

Heindel, Susanne *Alternate Doublecloth*

Holz, Matthias *Trees*

Honenbüchler, Heidemarie *Buttoned Boots*

Isaia, Katie *Graffiti*

Khanna-Ravich, Sheetal *Cypress Trees, Rooftops, Sheetal's Leaves*

Knudson, Bethanne *Let Go*

Korde, Pradyna *Pompoms*

Kovacs, Rudy *Adam and Eve III*

Lampinen, Tuulia *Cosmetica, Pine Motif, Sticks, Tiefly*

Lantz, Lindsay *Florence Map*

Lauri, Pirita *Afterparty*

Leoni, Alberto *Drago d'Oro*

Leoni, Francesca *Green Lancé*

Lessman-Moss, Janice *Red July: 336*

Loermans, Helena *Clipped Squares, Damask Squares, I Ching I, I Ching II, Pick-up-Sticks I, Pick-up-Sticks II, Puzzle Velvet, Shaded Squares*

Macali, Heather *Ogee*

McGruther Alaoui, Joy *Boxes*

Mokad, Ulrikka *Big Vermicelli, Vermicello*

Moor, Tina *Gondola, Gondola II*

Muller, Robin *Frost, Sandscroll, Skate Marks*

Ng, Elaine Yan Ling *SuperluxI, SuperluxII*

Nomura, Kazuyo *Lines, Vatten Väv*

Øestergaard, Stine *Bobles*

Olde, Melanie *Happles, Leaf, Melanie's Vermicelli*

Panitchpakdi, Narin *Copper, Narin's Velvet, Swirls*

Paul, Jan *Piazza San Marco—Day*

Peers, Simon *Simon's Fish*

Peters, Martine *Martine*

Porter, Martha *Honeycomb, Lemon Leaves*

Robertson, Jennifer *Banksia*

Sala Francesco *5 Weft Lampas, Yellow Roses*

Sargent, Zoe, LFS *Florence Map*

Schlechte, Eva *Chameleon*

Schoiswohl, Eveline *Himalaya*

Schryen, Annette *Triangles*

Shawcroft, Barbara *Briciola at Night*

Stucky, Antoinette *Small Squares*

Tanja, Valta *Morning*

Thomsson, Hans *Hans' Velvet, Onions, Paving Stones*

Thorn, Sara *Lace*

Tibbits, Veronica *Fish*

Tirabasso, Daniela *Uccelli*

Tritthart, Elizabeth *Blue Boxes, Gingko*

Verrelst, Lut *Handwash Only, Poppies*

Von Weissenberg, Janina *Leaf Skeleton*

Yate, Hannah *Cups and Spoons*

Ziek, Bhakti *Bhakti's Birds, Genesis, Primordial*

Collections

CJH (Author's collection)—Contemporary and historical textiles representative of different techniques

Cleveland Museum of Art—Weft-faced compound twill, silk, Sogdiana, eighth century C.E.

Cora Ginsburg LLC—European figured textiles, sixteenth to eighteenth centuries

John Marshall—Japanese figured silk velvet, lacquered and gilt-paper figured silk

LFAS (Lisio Foundation Archivio Storico)— *Santa Maria del Fiore*

LFCHT (Lisio Foundation Collection of Historic Textiles)—Warp-patterned plain weave, striped damask

Societé Le Manach—European figured textiles, eighteenth and nineteenth centuries

Photography and Illustrations

All photographs, artwork, and technical drawings belonging to individual artists are labeled with the artist's initials in the captions.

Figure 4.1 is the property of the Cleveland Museum of Art.

Figure 9.14 is the property of Paola Oudman.

All other illustrative materials—drawings, drafts and photographs—are the property of Dario Bartolini.

Index